温州市瓯海职业中专集团学校校本教材

电梯英语
Elevator English For Vocational School Students

主 编 王君

其他编委 陈碎芝 陈小敏 侯宝帅 李锋 林锦辉（排名不分先后）

浙江工商大學出版社

图书在版编目(CIP)数据

电梯英语 / 王君主编. —杭州：浙江工商大学出版社，2018.8

ISBN 978-7-5178-2939-3

Ⅰ.①电… Ⅱ.①王… Ⅲ.①电梯－英语－中等专业学校－教材 Ⅳ.①TU857

中国版本图书馆 CIP 数据核字(2018)第 203089 号

电梯英语

王 君 主编

责任编辑	柳 河
封面设计	林朦朦
责任印制	包建辉
出版发行	浙江工商大学出版社
	(杭州市教工路 198 号 邮政编码 310012)
	(e-mail:zjgsupress@163.com)
	(网址:http://www.zjgsupress.com)
	电话:0571-81902072,88831806(传真)
排　　版	杭州朝曦图文设计有限公司
印　　刷	虎彩印艺股份有限公司
开　　本	710mm×1000mm 1/16
印　　张	8
字　　数	143 千
版 印 次	2018 年 8 月第 1 版　2018 年 8 月第 1 次印刷
书　　号	ISBN 978-7-5178-2939-3
定　　价	32.00 元

版权所有　翻印必究　印装差错　负责调换

浙江工商大学出版社营销部邮购电话　0571-88904970

前　言

目前中国已成为世界上最大的电梯市场,整个电梯行业的发展蒸蒸日上,前景广阔。随着世界经济一体化步伐的加快,我国电梯制造与营销业也飞速发展,要求从事电梯行业的工作人员不但要精通本行业的业务技能,还应掌握一定的行业英语知识,才能更好、更全面地服务于电梯行业。因此,编写合适的电梯英语教材十分必要。

一、编写思路

本书以电梯技术的发展和应用为背景,从电梯行业各类人员的需求和实际工作要求着手,立足中等职业学校学生的特点,尤其考虑电梯专业学生的英语水平、学习特点,以电梯专业学生和电梯从业人员的实用和够用为原则进行编写。本书囊括了电梯基本知识、电梯机械系统知识、电梯安装、故障检测和电梯维保五大模块知识,对电梯专业涉及领域和知识归纳总结,每个模块的内容安排合乎电梯学习课程架构。本书每章节以图片学习为切入点,激发学习兴趣;以图片情境为载体,引发专业词汇学习,操练专业口语;以短文为范文,促进专业知识的学习;以鉴赏为模式,促进电梯文化的学习。

二、教材特色

1.实用性。翻阅本教材你能体会到的是:电梯专业最常见的词汇、最常见的电梯英文标识、最常见的电梯实用英语句子是本教材的三大关键词。电梯英语学习从词汇开始,到句子,到篇章,达到浅入、易入的目的。本书以培养中等职业学校电梯专业应用型人才为目的,贴近学生就业和企业用人的需求,使学生掌握工作所需的电梯专业英语知识、专业术语和词汇,并具备一定的专业英语阅读能力。

2.拓展性。本教材涵盖的不仅仅是电梯专业的基础知识,还包括电梯发展史、电梯文化、电梯行业的专业发展等相关内容。本书可作为中等职业学校、技工学校以及各类培训学校电梯销售、维修与保养等专业的教材。

由于编者水平有限,难免存在不当和错误之处,恳请广大读者批评指正。

CONTENTS

Chapter 1　Elevator Knowledge 1 ………………………………… 001

Section A　Warming Up ……………………………………… 002
Section B　Useful Sentences ………………………………… 005
Section C　Passages …………………………………………… 008
Section D　Elevator Culture ………………………………… 011

Chapter 2　Elevator Knowledge 2 ………………………………… 017

Section A　Warming Up ……………………………………… 018
Section B　Useful Sentences ………………………………… 021
Section C　Passages …………………………………………… 024
Section D　Elevator Culture ………………………………… 027

Chapter 3　Elevator Mechanical System ………………………… 035

Section A　Warming Up ……………………………………… 036
Section B　Useful Sentences ………………………………… 039
Section C　Passages …………………………………………… 042
Section D　Elevator Culture ………………………………… 046

Chapter 4　Elevator Installation ………………………………… 052

Section A　Warming Up ……………………………………… 053
Section B　Useful Sentences ………………………………… 056

Section C　Passages ········· 059
Section D　Elevator Culture ········· 062

Chapter 5　Elevator Test and Inspection ········· 069

Section A　Warming Up ········· 070
Section B　Useful Sentences ········· 073
Section C　Passages ········· 075
Section D　Elevator Culture ········· 079

Chapter 6　Troubleshooting and Maintenance ········· 086

Section A　Warming Up ········· 087
Section B　Useful Sentences ········· 090
Section C　Passages ········· 093
Section D　Elevator Culture ········· 097

Appendix 1　International Phonetics Alphabet ········· 111

Appendix 2　Keys to Exercises ········· 112

Appendix 3　Vocabulary ········· 117

Chapter 1　Elevator Knowledge 1

Section A Warming Up

1A: Match and say. 根据所给的单词和图片，配对并说出相应的名称。

elevator/lift	escalator
low speed elevator	mid-speed elevator
high speed elevator（express lift）	super high speed elevator

_____   _____

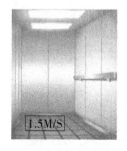

_____   _____

_____   _____

1B: Choose and read. 选择下列电梯标识的含义并读一读。

Take an elevator	No overload
Go down in an elevator	No leaning on the door
No playing in an elevator	Don't use in case of danger
Stand on right	In case of falling

_____

_____

_____

_____

_____

_____

_____

_____

1C: The most famous elevator companies in the world. 识别下列电梯公司标识。

_____

_____

_____

_____

_____

_____

_____

_____

1D: Word Bank. 词汇表。

elevator /ˈelɪveɪtə/ n. 电梯;升降机	lift /lɪft/ n. 电梯;升降机
escalator /ˈeskəleɪtə/ n. 自动扶梯	lean /liːn/ v. 斜靠
low speed elevator 低速电梯	mid-speed elevator 中速电梯
high speed elevator 高速电梯	express lift 高速电梯
super high speed elevator 超高速电梯	take an elevator 乘坐电梯
in case of danger 危急情况	

Section B Useful Sentences

2A: Match and write sentences. 根据图片,选写合适的句子。

> A. Please take the elevator to the top floor.
> B. He operated the lift and it started off.
> C. There's no elevator and we have to climb the stairs.
> D. The lift has a rated capacity of 14 persons.
> E. Every modern city is full of skyscrapers.
> F. In 1852, Elisha Graves Otis invented the first safety brake for elevator.

1.  _____

2.  _____

3.  _____

4.  _____

5. _____

6. _____

2B: Look and complete. 看图片信息，完成句子。

1. He _____ to the 5th floor.

2. —What is _____ of the elevator?
   —_____ persons.

3. The elevator _____, then stopped at Floor 12.

4. —Is there _____?
   —No, we have to _____.

5.

The _____ is beside the _____.

6.

Every modern city wants to build many _____ ____ to get the eyeballs of the world.

2C: Word Bank. 词汇表。

operate /ˈɒpəreɪt/ v. 操作；运转	stair /steə/ n. 楼梯
skyscraper /ˈskaɪskreɪpə/ n. 摩天大楼	brake /breɪk/ n. 刹车
rate /reɪt/ n. 比率	capacity /kəˈpæsəti/ n. 载容量
safety brake 紧急刹车，安全制动	top floor 顶楼
start off 开始	rated capacity 额定载容量
Elisha Graves Otis 伊莱沙·格雷夫斯·奥的斯	
be full of 充满	

Section C　Passages

3A：Read the dialogue and finish the exercise. 阅读对话，完成练习。

Meeting in an Elevator of the Company

(*Zhang Ming and Judy meet in a staff elevator in the company. Z=Zhang Ming, J=Judy.*)

J：Hold the door, please. …Thank you.

Z：Hi, Judy. The elevator is going up. Which floor?

J：Hi, Mr. Zhang, Floor 25 please.

Z：OK. I'm going to 28. By the way, how is your work?

J：Fine. How about you?

Z：Busy, but it's good busy.

J：That's good. Oh, this is my floor.

Z：Bye, Judy.

J：Bye, Mr. Zhang.

Word Bank. 词汇表。

---
staff /stɑːf/ *n.* 全体职员　　　　　company /ˈkʌmpəni/ *n.* 公司

hold the door 别关门

Fill in the form. 填写表格。

Meeting in an Elevator of the Company	
Zhang Ming's floor	
Judy's floor	
Zhang Ming's work	
Judy's work	
Zhang and Judy's relationship	

3B: Read the passage and finish the exercise. 阅读文章，完成练习。

The Only Way Is Up

Think of a modern city. It is full of great buildings. Every city wants to build skyscrapers. Most people live in cities. That means building upwards. But people didn't want to climb a mountain to work or home.

Elisha Otis, a US inventor, was the man who brought us the lift or elevator. In 1852, Otis invented the 1st safety brake for elevator and started the elevator history. However, most of his technology is very old. It had the same pulley system of the Pyramids of Egypt. Otis attached the system to a steam engine and developed the elevator brake, which stops the lift falling if the cords are broken. It was this invention that gained public confidence.

Today going in a lift is such an everyday thing that it would just be boring. But lift is one of the best places for people to get close to others and know each other better.

Word Bank. 词汇表。

upwards /ˈʌpwədz/ adv. 向上	technology /tekˈnɒlədʒi/ n. 科技
pulley /ˈpʊli/ n. 滑轮；滑车	system /ˈsɪstəm/ n. 系统；制度
attach /əˈtætʃ/ v. 贴；装；连接	fall /fɔːl/ v. 落下；跌落
cord /kɔːd/ n. 粗线；绳	confidence /ˈkɒnfɪdəns/ n. 信心
invention /ɪnˈvenʃn/ n. 发明	pulley system 滑轮系统
The Pyramids of Egypt 埃及金字塔	steam engine 蒸汽发动机

Notes. 注释。

1. That means building upwards. 那就意味着房子越建越高。
2. In 1852, Otis invented the 1st safety brake for elevator and started the elevator history. 1852年,奥的斯发明了第一个电梯安全制动器,开创了电梯的历史。
3. Otis attached the system to a steam engine and developed the elevator brake, which stops the lift falling if the cords are broken. 奥的斯把滑轮附在蒸汽发动机上,制成了电梯制动器,若绳索断裂,它能防止电梯坠落。
4. It was this invention that gained public confidence. 正是这个发明赢得了公众的自信心。
5. Today going in a lift is such an everyday thing that it would just be boring. 如今,乘坐电梯是如此平常的一件事,让人觉得厌倦。

Answer the following questions. 回答以下问题。

1. Is Elisha Otis an American or Englishman?

2. What did Otis invent for elevator in 1852?

3. What is the function of elevator brake system?

4. Otis's invention gained public confidence, didn't it?

Section D　Elevator Culture

4A：Magic Elevators. 神奇的电梯。

Bailong Elevator in China

This magic 326-metre-high elevator takes you up the side of one of the many dangerous and huge cliffs in Zhangjiajie, China. It is said to be the highest and heaviest outdoor elevator in the world. The Bailong Elevator has set three Guinness World Records—world's tallest full-exposure outdoor elevator, world's tallest double-deck sightseeing elevator and world's fastest passenger traffic elevator with biggest carrying capacity. However, the future of this elevator is uncertain as officials say that the elevator is causing problems of environment.

Word Bank. 词汇表。

magic /ˈmædʒɪk/ *adj.* 有魔力的	cliff /klɪf/ *n.* 悬崖，峭壁
uncertain /ʌnˈsɜːtn/ *adj.* 不确定的	environment /ɪnˈvaɪrənmənt/ *n.* 环境
full-exposure 全暴露的	outdoor elevator 户外电梯
Guinness World Record 吉尼斯世界纪录	

Notes. 注释。

1. This magic 326-metre-high elevator takes you up the side of one of the many dangerous and huge cliffs in Zhangjiajie, China. 在中国张家界，这座 326 米高的神奇的电梯把你从危险而巨大的悬崖侧面送上顶峰。
2. World's tallest full-exposure outdoor elevator, world's tallest double-deck sightseeing elevator and world's fastest passenger traffic elevator with biggest carrying capacity. 世界上最高的全暴露户外电梯，世界上最高的双层观光电梯和世界上速度最快、载容量最大的游客电梯。

Translation. 参考译文。

中国白龙电梯

在中国张家界,这座 326 米高的神奇的电梯把你从危险而巨大的悬崖侧面送上顶峰。据说,这是全世界最高、最重的户外电梯。白龙电梯创造了三项吉尼斯世界纪录:世界上最高的全暴露户外电梯,世界上最高的双层观光电梯和世界上速度最快、载容量最大的游客电梯。然而,正如政府官员所说的,这座电梯的未来还不确定,因为它会导致环境问题。

4B: More About Elevator. 电梯知识。

电梯发展简史
A Brief History of Elevator
(Part One)

电梯是垂直运行的电梯(通常也简称为电梯)、倾斜方向运行的自动扶梯、倾斜或水平方向运行的自动人行道的总称。据估计,截至 2002 年,全球在用电梯约 635 万台,其中垂直电梯约 610 万台,自动扶梯和自动人行道约 25 万台。电梯已成为人类现代生活中广泛使用的人员运输工具。人们对电梯安全性、高效性、舒适性的不断追求,推动了电梯技术的持续进步。

电梯的前身是采用人力、畜力和水力来拉升重量的垂直升降装置,这种装置在早期的农业社会就已经出现。据 Elevator History 记载,有凭有据的记录出现在公元前三世纪。根据 Landmark Elevator 中的描述,古希腊数学家、物理学家、科学家阿基米德是已知的首位电梯发明者。据 Otis World Wide 描述,他的设备由绳索和滑轮组成,绳子通过绞盘和杠杆缠绕在卷筒上。公元前 100 年前后,我国古人发明了辘轳。它采用卷筒的回转运动完成升降动作,因而增加了提升物品的高度。公元 80 年,角斗士和野生动物能够乘坐原始的升降机到达罗马大剧场中竞技场的高度。

早期的升降装置广泛采用人力作为拉升动力

中世纪，为给孤立地点运送人和物品的拉升升降装置开始大量出现，其中最著名的是位于希腊圣巴拉姆修道院的升降机。这个修道院位于距离地面大约 61 米（200 英尺）高的山顶上，人与货物上下的唯一途径就是升降机。

1405 年德国工程师 Konrad Kyeser 设计的电梯

从 18 世纪开始，机械力开始被用于升降机的运动。国王路易十五是最早将电梯设计成乘客专用的人，也就是我们所熟知的"飞行的椅子"。据 This is Versailles 描述，1743 年 Blaise-Henri Arnoult 将这款电梯安装在了凡尔赛宫。1833 年，一种使用往复杆的升降系统在德国哈尔茨山脉地区运送矿工。1835 年，一种被称为"绞盘机"的用皮带牵引的升降机出现在英国的一家工厂中。1846 年，第一部工业用水压式升降机出现。随着机器和工程技术的提高，其他动力的升降装置紧跟着出现了。

19 世纪初，欧美国家开始利用蒸汽机作为升降工具的动力。1845 年，威廉·汤姆逊研制出一台液压驱动的升降机，其液压驱动的介质是水。尽管升降工具被一代代富有革新精神的工程师们进行不断改进，但是被工业界普遍认可的升降机仍未出现，直到 1852 年世界第一台安全升降机诞生。

伊莱沙·格雷夫斯·奥的斯

1853 年，Elisha Otis 在 Crystal Palace 展示他的安全电梯

1852年，美国人伊莱沙·格雷夫斯·奥的斯发明了世界上第一台安全升降机。Elisha Graves Otis 首次将安全系统引入电梯内，它能够防止电梯因缆绳断裂而发生的垂直速降。据 Funding Universe Company Histories 的描述，当电梯缆绳断裂导致弹簧失去弹性的时候，锯齿状的棘轮能够将电梯稳定在当前位置。据 Columbia Elevator 电梯公司描述，1857年纽约百货商店首次为载客电梯站装了 Otis 的安全装置。

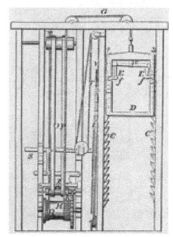

世界上第一台安全升降机的原理图

1853年9月20日，在纽约扬克斯，奥的斯在一家破产公司的部分场地上开办自己的电梯生产车间，奥的斯电梯公司由此诞生。

电梯制造业在纽约扬克斯诞生

1854年，在纽约水晶宫举行的博览会上，奥的斯第一次向公众展示了他的发明。自此，搭乘电梯不再被认为是"冒险者的游戏"。

1857年3月23日，奥的斯公司为地处纽约百老汇和布洛姆大街的 E. V. Haughwout 公司一座专营法国瓷器和玻璃器皿的商店安装了世界上第一台客运升降机。升降机由建筑物内的蒸汽动力站通过轴和皮带驱动升降机运动。

1857年，奥的斯公司第一台客运升降机安装在纽约 E. V. Haughwout 公司的一座商店

1862年，奥的斯公司采用单独蒸汽机控制的升降机问世。

1874年，罗伯特·辛德勒创建迅达公司。

1878年，奥的斯公司在纽约百老汇大街155号安装了第一台水压式乘客升降机，提升高度达34m。

据西门子公司的记载，1880年 Werner von Siemens 建造了第一台真正意义上用电的电梯。这款电梯由安装在底部平台的马达提供动力，通过一组基于电动机原理的齿轮组达到上升的效果。最初这款电梯是在 Mannheim Pfalzgau 农贸展览会上为首相设计的，但是却延期了两个月。最终电梯被证明是巨大的成功，成千上万的乘客都能受益。

1889年，世界第一个超高建筑电梯安装项目在法国巴黎完成。奥的斯公司在高度为324米的埃菲尔（Eiffel）铁塔中成功安装了升降电梯。按照铁塔底角的斜度及曲率，电梯在部分行程中须在倾斜的导轨上运行。

同年12月，奥的斯公司在纽约第玛瑞斯特大楼成功安装了一台直接连接式升降机。这是世界第一台由直流电动机提供动力的电力驱动升降机，从此诞生了名副其实的电梯。

奥的斯公司第一台电力驱动升降机

4C: Evaluation. 回顾所学知识,写下你已掌握的单词和句子。

Write down the words that you think are the most useful:

1. _____
2. _____
3. _____
4. _____
5. _____
6. _____
7. _____
8. _____

List the sentences you that think are the most important:

1. _____
2. _____
3. _____
4. _____
5. _____
6. _____
7. _____
8. _____

Chapter 2   Elevator Knowledge 2

Section A Warming Up

1A: Match and say. 根据所给单词和图片，配对并说出相应的名称。

passenger elevator	sightseeing lift
hospital elevator	freight elevator
residential lift	motor lift

_____   _____

_____   _____

_____   _____

1B: Guess and say. 识别电梯部件。

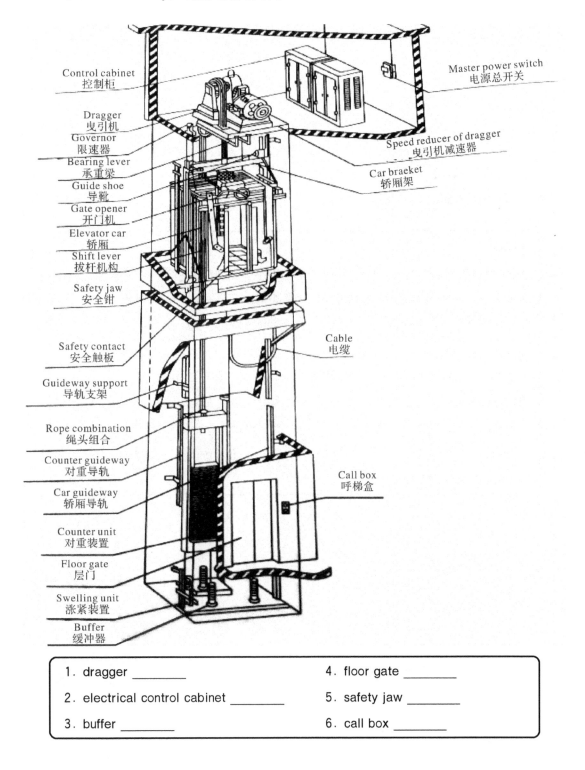

1. dragger _____
2. electrical control cabinet _____
3. buffer _____
4. floor gate _____
5. safety jaw _____
6. call box _____

1C：Match and say. 将电梯部件的中英文连线。

A. conveyor belt   1. 楼层指示器

B. speed reducer   2. 对重装置

C. floor indicator   3. 限速器

D. counter unit   4. 减速器

E. elevator cable   5. 电梯电缆

F. governor   6. 传送带

1D：Word Bank. 词汇表。

---

freight /freɪt/ n. （海运、空运或陆运的）货物

sightseeing /ˈsaɪtsiːɪŋ/ n. 游览；观光

motor /ˈməʊtə/ n. 汽车；发动机

residential /ˌrezɪˈdenʃl/ adj. 适合居住的

dragger /ˈdrægə/ n. 曳引机

buffer /ˈbʌfə/ n. 缓冲器

safety /ˈseɪfti/ n. 安全；平安

governor /ˈɡʌvənə/ n. 限速器

passenger elevator 乘客电梯

electrical control cabinet 电子控制柜

call box 呼梯盒

floor gate 层门

safety jaw 安全钳

conveyor belt 传送带

speed reducer 减速器

counter unit 对重装置

floor indicator 楼层指示器

elevator cable 电梯电缆

Section B Useful Sentences

2A: Match and write sentences. 根据图片,选写合适的句子。

> A. It is best for you to relax and enjoy in an elevator.
> B. A passenger elevator is designed to move people between a building's floors.
> C. Vehicle lifts are used within buildings or areas with limited place to move cars into the parking garage.
> D. A freight elevator is larger and has a greater capacity of carrying heavier loads from 2,500kg to 4,500kg.
> E. The Boeing 747 has a passenger double-deck aircraft lift.
> F. The elevator technology developed to provide the passenger and freight elevator in use today.

1.  _____

2.  _____

3.  _____

电 梯 英 语 | *Elevator English For Vocational School Students*

4. _____

5. _____

6. _____

2B: Look and complete. 看图片完成句子。

1. A _____ is _____ to move people between a _____ .

2. It is best for you to _____ and _____ in an elevator.

3. A _____ is larger and has a greater _____ of carrying heavier _____ from 2,500kg to 4,500kg.

4. The Boeing 747 has a _____ .

5.
Vehicle lifts are used within _____ or areas with limited place to _____ cars into the _____.

6.
The elevator technology developed to provide _____ _____ in use today.

2C: Word Bank. 词汇表。

relax /rɪˈlæks/ v. 放松
design /dɪˈzaɪn/ v. 设计
capacity /kəˈpæsəti/ n. 容量
provide...in use 投入使用
residential building 住宅楼
passenger double-deck aircraft lift 乘客双层飞机电梯
develop /dɪˈveləp/ v. 发展
vehicle /ˈviːəkl/ n. 交通工具；车辆
load /ləʊd/ n. 负载；载重量
parking garage 停车场
Boeing 747 波音 747 飞机

Section C Passages

3A: Read the passage and finish the exercise. 阅读文章,完成练习。

Enjoy Taking An Elevator

With the invention of the elevator, came high buildings and skyscrapers. Some people find elevators make them a little uncomfortable because there are many people standing close in a small room.

It is best for you to relax and enjoy in an elevator. When the elevator doors open, stand aside and let everyone out before you get in. Don't stare at people or stand too close, or others may feel nervous. If you want to get out of the lift, say "Excuse me." or "I'm sorry." If you stand close to the call buttons, ask others which floor they go to and push the buttons for them. This is a good manner. In the case of an emergency, follow the instructions inside the elevator and try to keep others calm.

Word Bank. 词汇表。

> aside /əˈsaɪd/ *adv.* 在旁边　　　　　　button /ˈbʌtn/ *n.* 按钮;纽扣
> nervous /ˈnɜːvəs/ *adj.* 紧张不安的　　　calm /kɑːm/ *adj.* 平静的;镇静的
> uncomfortable /ʌnˈkʌmftəbl/ *adj.* 使人不舒服的
> call buttons 呼梯按钮　　　　　　　　　stare at 盯着看
> in the case of an emergency 紧急情况时　follow the instructions 按照指令

Notes. 注释。

1. With the invention of the elevator, came high buildings and skyscrapers. 随着电梯的发明,出现了高楼和摩天大厦。

2. Some people find elevators make them a little uncomfortable because there are many people standing close in a small room. 有些人认为电梯会令他们感到有点不舒服,因为很多人近距离地站在一个小空间里。

3. If you stand close to the call buttons, ask others which floor they go to and push the buttons for them. This is a good manner. 如果你站在呼梯按钮边上,问别人到哪层楼,为他们按下按钮,这是有礼貌的表现。

4. In the case of an emergency, follow the instructions inside the elevator and try to keep others calm. 紧急情况时,遵循电梯里的指示,努力让其他乘客保持平静。

Decide the sentences T (true) or F (false). 判断句子正误。

1. Most people find elevator makes them a little uncomfortable. ( )
2. Before you get in a lift, let others out first. ( )
3. It is polite to stand close to other people when taking a lift. ( )
4. It is a good manner to push the buttons for other people. ( )
5. Keeping all people calm is important in the case of an emergency. ( )

3B: Read the passage and finish the exercise. 阅读文章,完成练习。

<h3 style="text-align:center">A Freight Elevator</h3>

A freight elevator is used to do just as its name: to carry freight or goods. It is designed to carry goods rather than people. There is often a notice in the car that passengers are not allowed though some do both to allow the operators and the loading goods along for the ride.

A freight elevator is usually larger and has a greater capacity of carrying heavier loads from 2,500kg to 4,500kg than a passenger elevator. Some can even handle as much as 100,000 pounds (45,359kg).

A freight elevator often has a manual door, and sometimes multiple doors.

Word Bank. 词汇表。

goods /gʊdz/ n. 货物
allow /əˈlaʊ/ v. 允许；许可
though /ðəʊ/ conj. 虽然；尽管
manual /ˈmænjuəl/ adj. 手动的
pound /paʊnd/ n. 英镑
notice /ˈnəʊtɪs/ n. 通知
operator /ˈɒpəreɪtə/ n. 操作者
handle /ˈhændl/ v. 操纵；搬运
multiple /ˈmʌltɪpl/ adj. 多种多样的

Notes. 注释。

1. It is designed to carry goods rather than people. 它的设计是为了运货而不是载人。
2. There is often a notice in the car that passengers are not allowed though some do both to allow the operators and the loading goods along for the ride. 轿厢里一般都有告示：乘客禁止使用货梯，虽然有的货梯允许操作人员和货物一起运送。
3. A freight elevator is usually larger and has a greater capacity of carrying heavier loads from 2,500kg to 4,500kg than a passenger elevator. 货物电梯通常比乘客电梯有更大的空间和载容量，载容量达 2500 千克到 4500 千克。

Fill in the form. 填写表格。

\	A Freight Elevator
Function	
Capacity	
Loads	
Door	

Section D Elevator Culture

4A：Magic Elevators. 神奇的电梯。

The Gateway Arch in Missouri

One of the "must sees" of St. Louis, Missouri, is the Gateway Arch. The Gateway Arch is the symbol of the city of St. Louis and it is 630 feet in height. To go to the top of the Arch, passengers in groups of five enter an "egg" room which has five seats and a flat floor. Eight rooms are linked to be a train. The most exciting part of visiting the Gateway Arch is the trip to the top of the Arch by a tram from either the south leg or the north leg of the Arch. The trip to the top of the Arch takes four minutes, and the trip back down takes three minutes. The car doors have narrow glass windows. Passengers can see the indoors stair and structure of the Arch during the trip. When reaching the observation area, you will have a magnificent view of the St. Louis.

Word Bank. 词汇表。

---

gateway /ˈgeɪtweɪ/ n. 门

symbol /ˈsɪmbl/ n. 象征

height /haɪt/ n. 高度

structure /ˈstrʌktʃə/ n. 结构；构造

magnificent /mægˈnɪfɪsnt/ adj. 壮丽的；宏伟的

must see 必看的

St. Louis, Missouri（美国）密苏里州圣路易斯市

arch /ɑːtʃ/ n. 拱门

link /lɪŋk/ v. 连接；联系

foot /fʊt/ n. 英尺（复数 feet）

Notes. 注释。

1. One of the "must sees" of St. Louis, Missouri, is the Gateway Arch. (美国)密苏里州圣路易斯市必看的景点之一就是拱门电梯。
2. To go to the top of the Arch, passengers in groups of five enter an "egg" room which has five seats and a flat floor. 想要到达电梯顶部,每5位乘客分为一组进入一个鸡蛋状的房间,房间是平底的,有五个座位。
3. The most exciting part of visiting the Gateway Arch is the trip to the top of the Arch by a tram from either the south leg or the north leg of the Arch. 参观拱门电梯最激动人心的是从拱门南面或北面乘电车到达电梯顶端。
4. When reaching the observation area, you will have a magnificent view of the St. Louis. 到达观景区,你将领略圣路易斯市的美妙风景。

Translation. 参考译文。

密苏里州拱门电梯

(美国)密苏里州圣路易斯市必看的景点之一就是拱门电梯。拱门电梯是圣路易斯城市的象征,它的高度是630英尺。想要到达电梯顶部,每5位乘客分为一组进入一个鸡蛋状的房间,房间是平底的,有五个座位。8个房间串在一起组成一节车厢。参观拱门电梯最激动人心的是从拱门南面或北面乘电车到达电梯顶端。上到拱门顶部需要4分钟而下来则是3分钟。房间的门都有狭窄的玻璃窗。旅途中,乘客们可以看见室内的楼梯以及拱门的结构。到达观景区后,你将领略圣路易斯市的美妙风景。

4B: More About Elevator. 电梯知识。

电梯发展简史
A Brief History of Elevator
(Part Two)

1891年,纽约企业家杰西·雷诺(Jesse Wilfred Reno)在美国科尼岛码头设计制造出世界上第一部自动扶梯,它当时被称为"倾斜升降机"。这种自动扶梯采用输送带原理,一条分节的坡道以20度至30度坡度移动。扶梯的起止点都有齿长40厘米的梳状铲,与脚踏板上的凹齿啮合。乘客站在倾斜移动的节片上,不必举足,便能上下扶梯。

1898年,美国设计者查尔斯·西伯格(Charles David Seeberger)买下了一项扶梯专利,并与奥的斯公司携手改进制作。

1899年7月9日,第一台奥的斯—西伯格梯阶式(梯级是水平的,踏板用硬木制成,有活动扶手和梳齿板)扶梯试制成功。这是世界第一台真正的扶梯。

1900年,奥的斯—西伯格梯阶式扶梯在法国巴黎国际博览会上展出并取得巨大成功。在这届博览会上,由杰西·雷诺(Jesse Wilfred Reno)设计的扶梯同样引人注目。在接下来的10年里,奥的斯—西伯格和雷诺是世界上仅有的扶梯生产竞争者。

同年,西伯格把拉丁文中"scala(楼梯)"一词与"elevator(电梯)"一词的字母结合,创造了"escalator(如今称为自动扶梯)"一词,并将其注册为产品商标。

1902年,瑞士迅达电梯公司开发了采用自动按钮控制的乘客电梯。

1903年,奥的斯公司在纽约安装了第一台直流无齿轮曳引电梯。

1907年,奥的斯公司在上海的汇中饭店(今和平饭店南楼)安装了两台电梯。这两台电梯被认为是我国最早使用的电梯。

1910年,西伯格根据合作协议把自己全部专利出售给奥的斯公司,其中就包括"escalator"这一注册商标。直到1950年,奥的斯公司一直拥有该商标的使用权。1950年,根据《商标法》的规定,"escalator"名称的专有权到期,成为扶梯的通称。

1911年,奥的斯收购雷诺。

随着建筑的不断发展,电梯需要到达更高的楼层,乘客也希望楼层之间的速度能够更快。

1915年,奥的斯公司设计了自动平层微动装置,首次用于美国海军舰队的电梯。

1920年,奥的斯公司把雷诺的倾斜板条式扶梯和西伯格的梯阶式扶梯重新进行设计,使扶梯性能大为改观。

1922年,奥的斯公司制造了世界上第一台现代化自动扶梯。这台扶梯采用水平楔槽式梯级与梳齿板相结合,这种设计方式后来被其他扶梯制造商广泛使用并一直沿用至今。

1924年,奥的斯公司在纽约新建的标准石油公司大楼安装了第一台信号控制电梯。这是一种自动化程度较高的有司机电梯。

第一台信号控制电梯

1926年,迅达公司开始生产采用沃德—伦纳德

(发电机—电动机组)系统驱动的直接曳引式电梯。

1931 年,奥的斯公司在纽约安装了世界第一台双层轿厢电梯。双层轿厢电梯增加了额定载重量,节省了井道空间,提高了输送能力。

1946 年,奥的斯公司设计了群控电梯。

1949 年,首批群控电梯落户纽约的联合国大厦。

1956 年,世界第一台交流驱动电梯在迅达公司诞生。

1967 年,奥的斯公司为美国纽约世界贸易中心大楼安装了 208 台电梯和 49 台自动扶梯。

1968 年,奥的斯公司为美国芝加哥的 Time-Life 大厦安装了一台双层轿厢电梯系统。

1974 年,奥的斯公司在荷兰阿姆斯特丹国际机场安装了 200 米长的自动人行道。这是当时欧洲最长的一条自动人行道。

1975 年,奥的斯电梯现身世界最高独立式建筑——加拿大多伦多 CN 电视塔。这座总高度达 553.34m 的电视塔内安装了四台奥的斯公司特制的玻璃围壁观光电梯。

1976 年 7 月,日本富士达公司开发出速度为 10.00m/s 的直流无齿轮曳引电梯。

1977 年,日本三菱电机公司开发出可控硅—伦纳德无齿轮曳引电梯。

1979 年,奥的斯公司开发出第一台基于微处理器的电梯控制系统 Elevonic101,从而使电梯电气控制进入一个崭新的发展时期。

1980 年,奥的斯公司发布了一个称为 Otis Plan 的计算机程序,它可以帮助建筑师们为新建或改造建筑物确定电梯的最佳形式、速度以及数量等配置方案。

1983 年,奥的斯公司开发出 OTIS LINE,它是一个基于计算机的全天候(24 小时)召修服务系统。

1983 年,三菱电机公司开发出世界第一台变压变频驱动的电梯。

1985年，三菱电机公司研制出曲线运行的螺旋型自动扶梯，并成功投入生产。螺旋型自动扶梯可以节省建筑空间，具有装饰艺术效果。

三菱电机公司的螺旋型自动扶梯

1988年2月，富士达公司将电梯群控管理系统"FLEX8800"系列商品化，这是一套应用模糊理论与人工智能技术的管理系统。

1988年，奥的斯公司发布了REM，它是一个可以远程监测电梯性能的计算机诊断系统。

1989年，奥的斯公司在日本发布了无机房线性电机驱动的电梯。

1990年，三菱电机公司首次将变频驱动系统应用于液压电梯。

1991年，三菱电机公司开发出带有中间水平段的大提升高度自动扶梯。这种多坡度型自动扶梯在大提升高度时可降低乘客对高度的恐惧感，并能与大楼楼梯结构协调配置。

1992年12月，奥的斯公司在日本东京附近的Narita机场安装了穿梭人员运输系统。穿梭轿厢悬浮于一个气垫上，运行速度可达9.00m/s，运行过程平滑、无声。后来，奥的斯公司又在奥地利、南非以及美国其他一些地区安装了该系统。

奥的斯公司的水平穿梭人员运输系统

1993年,三菱电机公司在日本横滨 Landmark 大厦安装了速度为 12.50m/s 的超高速乘客电梯,这是当时世界上速度最快的乘客电梯。

1993年,据《日立评论》报道,日本日立制作所开发出可以乘运大型轮椅的自动扶梯,这种扶梯的几个相邻梯级可以联动形成支持轮椅的平台。

1995年,奥的斯公司引入 REMⅢ,它是奥的斯最先进的远程电梯监控系统。电梯远程监控系统逐渐成为各大公司提高服务的手段。

1995年,三菱电机公司开发出 MEL ART 全彩色图形喷漆技术,用于电梯部件(如电梯门)的喷漆。

1996年3月,芬兰通力电梯公司发布无机房电梯系统,电机固定在机房顶部侧面的导轨上,由钢丝绳传动牵引轿厢。整套系统采用永磁同步电机变压变频驱动。

芬兰通力电梯公司的无机房电梯系统

1996年,奥的斯公司推出"Odyssey",这是一个集垂直运输与水平运输的复合运输系统。该系统采用直线电机驱动,在一个井道内设置多台轿厢,轿厢在计算机导航系统控制下,能够在轨道网络内交换各自运行路线。

奥的斯公司推出的"Odyssey"

同年，迅达电梯公司推出"Miconic10"目的楼层厅站登记系统。该系统操纵盘设置在各层站候梯厅，乘客在呼梯时只需登记目的楼层号码，就会知道最佳的乘梯方案，从而提前去该电梯厅门等候。待乘客进入轿厢后不再需要选层。

1996年，迅达电梯公司推出的"Miconic10"目的楼层厅站登记系统

1996年，三菱电机公司开发出采用永磁电机无齿轮曳引机和双盘式制动系统的双层轿厢高速电梯，并安装于上海的Mori大厦。

1997年4月，迅达电梯公司在慕尼黑展示了无机房电梯，该电梯无需曳引绳和承载井道，自驱动轿厢在自支撑的铝制导轨上垂直运行。

1997年，通力电梯公司在芬兰建造了当今世界上行程最大（350m）的地下电梯试验井道，实际最大行程330m，理论上可测试速度为17.00m/s的电梯。

1999年，奥的斯公司发布3个电子商务产品：电子直销e＊Direct、电子服务e＊Service与电子显示e＊Display。电子显示e＊Display是通过电梯轿厢内的一块平板显示屏，向乘客提供新闻、天气预报、股市行情、体育比赛结果等信息，同时也可提供楼层指南，发布广告。

20世纪90年代末，富士达公司开发出变速式自动人行道。这种自动人行道以分段速度运行，乘客从低速段进入，然后进入高速平稳运行段，再进入低速段离开。这提高了乘客上下自动人行道时的安全性，也缩短了长行程时的乘梯时间。

4C: Evaluation. 回顾所学知识，写下你已掌握的单词和句子。

Write down the words that you think are the most useful:

1. _____
2. _____
3. _____
4. _____
5. _____
6. _____
7. _____
8. _____

List the sentences that you think are the most important:

1. _____
2. _____
3. _____
4. _____
5. _____
6. _____
7. _____
8. _____

Chapter 3　Elevator Mechanical System

Section A Warming Up

1A: Match and say. 根据所给单词和图片,配对并说出相应的名称。

counterweight	pulley
elevator shaft/well	sensor
wire rope	guide shoe

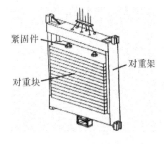

_____

_____

_____

_____

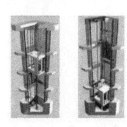

_____

_____

| Chapter 3 | Elevator Mechanical System

1B: Match and read. 将下列的中英文连线。

brake system   电梯机械系统
electronic control system   电气控制系统
elevator mechanical system 制动器系统
rotational speed   额定负荷
balance   安全系统
rated load   转速
safety system   平衡
elevator computer   电梯控制板

1C: Guess and say. 识别电梯部件。

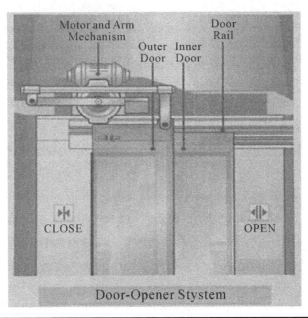

door(gate) opener system _____   motor _____
door rail _____   outer door _____
inner door _____

1D: Word Bank. 词汇表。

counterweight /ˈkaʊntəweɪt/ n. 对重	shaft /ʃɑːft/ n. 机井
sensor /ˈsensə/ n. 传感器	balance /ˈbæləns/ n. 平衡
motor /ˈməʊtə/ n. 发动机；马达，汽车	rail /reɪl/ n. 轨道
wire rope 钢丝绳	guide shoe 导靴
electric motor 电动机	brake system 制动器系统
rotational speed 转速	safety system 安全系统
elevator well 电梯机井	elevator computer 电梯控制板
electronic control system 电气控制系统	
elevator mechanical system 电梯机械系统	
door opener system 开门机系统	outer door 层门，厅门
inner door 轿门	door rail 门导轨

Section B Useful Sentences

2A: Match and write sentences. 根据图片,选写合适的句子。

> A. An electric motor hoists the cars up and down, including a brake system.
> B. The counterweights balance the cars. It makes the motor easier to raise and lower the car.
> C. Elevator brake is an important safety device. It is one of the factors to ensure the safe operation of the lift.
> D. The doors on the cars are operated by an electric motor which is hooked up to the elevator computer.
> E. Many elevators have a motion sensor system that stops the doors closing if somebody is between them.
> F. Safety systems protect the passengers from falling into the bottom of the shaft.

1.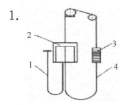

   _____

2.

   _____

3.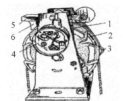

   _____

4. _____

5. _____

6. _____

2B: Look and complete. 看图片信息完成句子。

1. The counterweights _____ the cars. It makes the motor easier to raise and lower the car.

2. An electric motor _____ the cars _____, including a brake system.

3. The doors on the cars are _____ by _____ which is hooked up to _____.

4. _____ is an important safety device. It is one of the factors to ensure the safe operation of _____.

5.

| Many elevators have a _____ system that stops the doors _____ if somebody is between them. |

6.

| Safety systems _____ the passengers from _____ the bottom of the shaft. |

2C: Word Bank. 词汇表。

hoist /hɔɪst/ v. 升起；吊起	factor /ˈfæktə/ n. 因素；要素
ensure /ɪnˈʃɔː; -ˈʃʊə/ v. 保证；担保	safety device 安全设备
electric motor 电马达	cut off power 切断电源
hooked up 钩住；提拉	motion sensor system 运行传感器系统
protect from doing sth. 保护……以免……	

Section C Passages

3A：Read the passage and finish the exercise. 阅读文章，完成练习。

What Is an Elevator?

From the viewpoint of someone who travels from the 1st floor to the 25th floor, an elevator is simply a metal box with doors that closes on one floor and then opens again on another. For those of us are much more. The key parts of an elevator are:

One or more cars metal boxes that rises up and down.

Counterweights that balance the cars. It makes it easier for the motors to raise and lower the cars—just like a see-saw. It makes it easier to lift someone's weight compared to lifting him in your arms.

An electric motor that hoists the cars up and down, including a brake system.

A system of strong metal cables and pulleys running between the cars and the motors.

Various safety systems to protect the passengers if a cable breaks.

Word Bank. 词汇表。

---

viewpoint /ˈvjuːpɔɪnt/ n. 观点；看法　　raise /reɪz/ v. 提升

lower /ˈləʊə/ v. 使……降下　　see-saw /ˈsiːsɔː/ n. 跷跷板

Notes. 注释。

1. From the viewpoint of someone who travels from the 1st floor to the 25th floor, an elevator is simply a metal box with doors that closes on one floor and then opens again on another. 在那些乘坐电梯从 1 楼到 25 楼的人看来，电梯只是个带

门的金属箱子,在某一层关门,再在另一层开门而已。

2. For those of us are much more. 对于我们来说,就不仅仅是这些了。

3. It makes it easier for the motors to raise and lower the cars—just like a see-saw. 它使得电机驱动轿厢上下运行更加方便,就像一个跷跷板。

4. It makes it easier to lift someone's weight compared to lifting him in your arms. 相比用手臂把某人整个举起而言,坐在跷跷板上更容易把人举起来。

5. Various safety systems to protect the passengers if a cable breaks. 在电缆破损时,电梯有用来保护乘客的各种安全系统。

Fill in the form. 填写表格。

What Is an Elevator?

Parts	Function
Cars metal boxes	
Counterweights	
Electric motor	
Metal cables and pulleys	
Safety systems	

3B: Read the passage and finish the exercise. 阅读文章,完成练习。

Elevator Safety System

Elevator brake is an important safety device. It is one of the factors to ensure the safe operation of a lift. The role of the brake is in two: (1) Make the car stop when the power is cut off. (2) When the elevator stops, the brake should ensure

125% of rated load to keep a still car and the location.

Elevators have several safety devices. Even if all the ropes were to break, or the pulley system were to release them, it is impossibly an elevator would fall to the bottom of the shaft. Roped elevator cars have brake systems or safeties that grab onto the rail when the car moves too fast.

The doors on the cars are operated by an electric motor, which is hooked up to the elevator computer. The computer orders the motor to open the doors when the car arrives at a floor and close the doors before the car starts moving again. Many elevators have a motion sensor system that stops the doors closing if somebody is between them.

Word Bank. 词汇表。

still /stɪl/ *adj.* 静止的；寂静的	location /ləʊˈkeɪʃn/ *n.* 地点；位置
release /rɪˈliːs/ *v.* 松开，释放	grab /græb/ *v.* 抓住；夺取
hook /hʊk/ *v.* （使）钩住；*n.* 钩子	

Notes. 注释。

1. When the elevator stops, the brake should ensure 125% of rated load to keep a still car and the location. 电梯停止时，制动器应该在125%的额定载容量下保证电梯轿厢静止不动且位置不变。

2. Even if all the ropes were to break, or the pulley system were to release them, it is impossibly an elevator would fall to the bottom of the shaft. 即使所有的钢丝绳都断了或者滑轮系统松开了，电梯也不可能坠落至机井底部。

3. Roped elevator cars have brake systems or safeties that grab onto the rail when the car moves too fast. 轿厢连着钢丝绳，一旦轿厢滑行太快，制动器或安全设备会将轿厢夹住在门轨道上。

Decide the sentences T (true) or F (false). 判断句子正误。

1. Elevator brake is an important safety device. ( )
2. When the elevator stops, the brake should ensure 100% of rated load to keep a

still car and the location. ( )
3. An elevator only has one safety device. ( )
4. If all the ropes were to break, an elevator would fall to the bottom of the shaft.
 ( )
5. Now all elevators have motion sensor systems. ( )

Section D Elevator Culture

4A: Magic Elevators. 神奇的电梯。

Luxor Inclinator Elevator in Nevada

In Las Vegas, Nevada, at the Luxor Hotel, is the Inclinator. The shape of this casino is a pyramid. Therefore, the elevator travels up the side of the pyramid at a 39 degree angle.

Although people refer to this "inclined elevator" as an inclinator, this is incorrect. An inclinator is a stair lift developed by Inclinator Company of America many years ago. Therefore, the Luxor installation is just Otis Elevator's version of a generic "Inclined Elevator".

Word Bank. 词汇表。

Las Vegas *n*. 拉斯维加斯(美国城市名)

Nevada *n*. 内华达州(美国州名)

casino /kəˈsiːnəʊ/ *n*. 赌场   version /ˈvɜːʃn/ *n*. 版本

> generic /dʒə'nerɪk/ adj. 通用的　　　　Luxor Hotel 卢克索酒店
> Luxor Inclinator Elevator 卢克索倾斜仪电梯
> 39 degree angle 39 度角

Notes. 注释。

1. In Las Vegas, Nevada, at the Luxor Hotel, is the Inclinator. 在美国内达华州的拉斯维加斯市的卢克索酒店,有个倾斜仪电梯。
2. Although people refer to this "inclined elevator" as an inclinator, this is incorrect. 虽然人们喜欢把这个倾斜的电梯叫作倾斜仪,但这是不对的。
3. Therefore, the Luxor installation is just Otis Elevator's version of a generic "Inclined Elevator". 因此,卢克索电梯只是奥的斯电梯公司通用的倾斜仪电梯的一个安装版本而已。

Translation. 参考译文。

内达华州卢克索倾斜仪电梯

在美国内达华州的拉斯维加斯市的卢克索酒店,有个倾斜仪电梯。拉斯维加斯赌场是金字塔形状。所以电梯沿着金字塔的一边呈 39 度角上行。

虽然人们喜欢把这个倾斜的电梯叫作倾斜仪,但这是不对的。倾斜仪电梯只是美国倾斜仪公司多年前开发的一款楼梯升降机。因此,卢克索电梯只是奥的斯电梯公司通用的倾斜仪电梯的一个安装版本而已。

4B: More About Elevator. 电梯知识。

电梯发展简史

A Brief History of Elevator

(Part Three)

2000 年 5 月,迅达电梯公司发布 Eurolift 无机房电梯。它采用高强度无钢丝绳芯的合成纤维曳引绳牵引轿厢。每根曳引绳大约由 30 万股细纤维组成,比传统钢丝绳轻 4 倍。绳中嵌入石墨纤维导体,能够监控曳引绳的轻微磨损等变化。

Eurolift 无机房电梯

2000年，奥的斯公司开发出Gen2无机房电梯。它采用扁平的钢丝绳加固胶带牵引轿厢。钢丝绳加固胶带外面包裹聚氨酯材料，柔性好。无齿轮曳引机呈细长形，体积小、易安装，耗能仅为传统齿轮传动机器的一半。该电梯运行不需润滑油，因此更具环保特性。Gen2无机房电梯成为业界公认的"绿色电梯"。

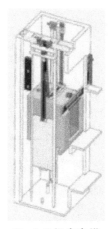

Gen2 无机房电梯

2002年4月17～20日，三菱电机公司在第五届中国国际电梯展览会上展出了倾斜段高速运行的自动扶梯模型，其倾斜段的速度是出入口水平段速度的1.5倍。该扶梯不仅能够缩短乘客的乘梯时间，同时也提高了乘客上下扶梯时的安全性与平稳性。

2003年2月，奥的斯公司发布新型的"NextStep"自动扶梯。它采用了革新的"Guarded"踏板设计，梯级踏板与围裙板成为协调运行的单一模块。它还采用了其他一些提高自动模块梯安全性的新技术。

2003年，正在建设中的台北国际金融中心大厦将安装速度为1010m/min（16.8m/s）的超高速电梯。该电梯由日本东芝电梯公司生产，提升高度达388m。

台北国际金融中心大厦

生活在继续,科技在发展,电梯也在进步。电梯的材质由黑白到彩色,样式由直式到斜式,在操纵控制方面更是步步出新——手柄开关操纵、按钮控制、信号控制、集选控制、人机对话等,多台电梯还出现了并联控制,智能群控。双层轿箱电梯展示出节省井道空间,提升运输能力的优势。变速式自动人行道扶梯大大节省了行人的时间,不同外形的扇形、三角形、半棱形、圆形观光电梯则使身处其中的乘客的视线不再封闭。

一个半世纪的风风雨雨,翻天覆地的是历史的变迁,永恒不变的是电梯提升现代人生活质量的承诺。

据美国有线电视新闻网络(CNN)报道,中国的建筑目前包揽了电梯的三项纪录:最快、最高和速度最快的双层电梯。上海中心大厦是世界上第二高的建筑,高达2074英尺(约合632米),它的电梯由日本三菱电机公司设计,到达121层的运行速度为每秒67英尺(约合20.5米每秒)。

上海中心大厦

然而这种创造最高和最好建筑的竞赛永远不会停歇。据 CNN 报道,将于 2019 年完工的沙特阿拉伯吉达塔届时将会是世界上最高的建筑,同时拥有世界上速度最快的电梯。对于高层建筑的电梯来说,为了保证其上升的高度和速度,需要进行多项审核。总部位于芬兰的 Kone 公司,他们的电梯运用了碳纤维绳索,强大的拉力能够将电梯送上 2165 英尺的高度(约合 660 米)。

为了让电梯能够上升的更高、更快,发明家们不断改善技术,并增添了许多新的安全特性。

例如,2009 年 Otis Elevator 公司的一群发明家就申请了电梯加速度和速度过快保护的相关专利。该系统一旦发现电梯超速,就会触发电磁制动器的刹车。来自瑞士的发明家 Juan Carlos Abad 于 2011 年申请了电梯安全电路的专利,该系统能够在电梯紧急刹车时控制轿厢并开始减速。

新的技术依旧在研发,电梯也会升得更高,运行得更快,更安全。

现在人们甚至想在电梯的缆绳中加入磁力。据 Business Insider 报道,德国 ThyssenKrupp 公司一直在研发 MULTI 电梯,这款电梯竟然运用了磁悬浮原理。今后的电梯不仅会大大减少占地面积,同时还会大幅提高人们的出行效率,多个电梯管道将会同时工作。就像电影《查理和巧克力工厂》中的电梯,它可以横纵穿越,创造各种新的可能性。

德国 ThyssenKrupp 公司研发的 MULTI 电梯

但是我们到底能造多高的电梯呢?科幻小说作家 Arthur C. Clarke 在他的作品《Fountains of Paradise》中称电梯可以将我们送入太空。在他的小说中,仅需五小时你就能从地球表面到达太空殖民地,而且你会看到"令你感到惊讶无比的景象"——周边都是从地球而来的乘客,地球在你的脚下越来越小。

据美国国家航空和宇宙航行局(NASA)所言,不久的将来能到达太空的电梯就会梦想成真。到那时,电梯将从一个 31 英里(约合 50 公里)的基地塔延伸至一颗距离地球 22236 英里(约合 35786 公里高)的地球同步卫星上。同时磁悬浮电梯车将会出现在 4~6 个轨道上以每小时几千公里的速度运行。普通人登上太空的梦想将实现。

4C:Evaluation. 回顾所学知识,写下你已掌握的单词和句子。

Write down the words that you think are the most useful:
1. _____
2. _____
3. _____
4. _____
5. _____
6. _____
7. _____
8. _____

List the sentences that you think are the most important:
1. _____
2. _____
3. _____
4. _____
5. _____
6. _____
7. _____
8. _____

Chapter 4　Elevator Installation

Section A Warming Up

1A: Match and say. 根据所给单词和图片，配对并说出相应的名称。

elevator car	lift motor room/penthouse
breaker	selector
speed reducer	elevator pit

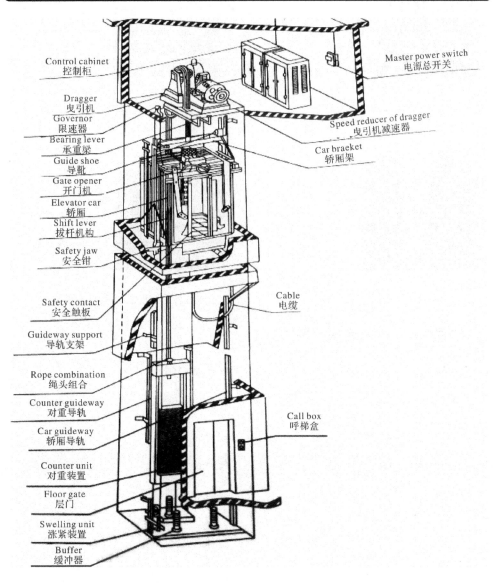

_____
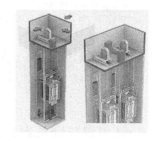
_____

_____

_____

_____

_____

1B: Match and say. 连线并朗读。

A. elevator man/worker   1. 电梯工程师

B. elevator engineer   2. 电梯工

C. elevator repairer   3. 电梯操作员

D. elevator operator   4. 电梯安装员

E. elevator installer   5. 电梯维修师

1C：Word Bank. 词汇表。

> breaker /ˈbreɪkə/ *n.* 断路器　　　　pit /pɪt/ *n.* 底坑
> installer /ɪnˈstɔːlə/ *n.* 安装员
> penthouse /ˈpenthaʊs/ *n.* 机房；阁楼套房
> selector /sɪˈlektə/ *n.* 选层器　　　engineer /ˌendʒɪˈnɪə/ *n.* 工程师
> repairer /rɪˈpeərə/ *n.* 修理员　　　speed reducer 减速器
> lift motor room 机房

Section B Useful Sentences

2A: Match and write sentences. 根据图片,选写合适的句子。

> A. The contactors, relays, resistors and transformers, etc., are in the control cabinet.
>
> B. Call boxes include car call boxes and landing call boxes. Car call and landing call are call signals.
>
> C. MRL installations need fewer materials and less work time: No well holes to be drilled, no pits to be waterproofed.
>
> D. For a faster and smoother installation, be sure you have all necessary parts and tools before the installation.
>
> E. Elevator installers and repairers usually major in installation, maintenance or repair work.
>
> F. After the last call, car will return to home landing if no calls are received. This is home landing parking.

1. 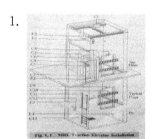 _____

2.  _____

| Chapter 4 | Elevator Installation

3.  _____

4.  _____

5.  _____

6.  _____

2B: Look and complete. 看图片信息完成句子。

1.  The contactors, relays, resistors and transformers, etc., are in _____.

2. Elevator installers and repairers, also called elevator constructors, usually major in _____, _____ or _____.

3. 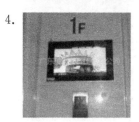 For a faster and smoother installation, be sure you have all necessary _____ and _____ before the _____.

4. 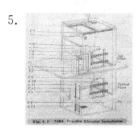 After the last call, car will return to home landing if no calls are received. This is _____.

5.  In MRL installations, there are no well holes to be _____, no pits to be _____, no machines to be hoisted to the penthouse floor.

6. Call boxes include _____ boxes and _____ boxes. Car call and landing call are call signals.

2C: Word Bank. 词汇表。

contactor /ˈkɒntæktə/ n. 触点	relay /ˈriːleɪ/ n. 继电器；中继设备
resistor /rɪˈzɪstə/ n. 电阻器	transformer /trænsˈfɔːmə/ n. 变压器
installation /ˌɪnstəˈleɪʃn/ n. 安装；设置	drill /drɪl/ n. 钻；钻机
waterproof /ˈwɔːtəpruːf/ adj. 防水的	maintenance /ˈmeɪntənəns/ n. 维护
landing call 层站停靠	car call 轿厢呼叫
major in 主攻，专业	home landing parking 基站停靠
MRL installation 无机房安装	

Section C Passages

3A: Read the passage and finish the exercise. 阅读文章,完成练习。

Work Environment of Elevator Installers

Elevator installers lift and carry heavy equipment and parts. They may work in narrow spaces or awkward positions. The dangers include falls, electrical shock, muscle strains, and other injuries related to handling heavy equipment. To prevent injuries, workers often wear hardhats, harnesses, ear plugs, safety glasses, protective clothing, shoes and masks. Because most of their work is in buildings, installers and repairers lose less work time because of bad weather than most other workers do in the construction trades.

Word Bank. 词汇表。

awkward /ˈɔːkwəd/ *adj.* 令人尴尬的	injury /ˈɪndʒəri/ *n.* 伤害;损害
hardhat /ˈhɑːdˈhæt/ *n.* 安全帽	harness /ˈhɑːnɪs/ *n.* 背带,保护带
plug /plʌɡ/ *n.* 塞子;插头	mask /mɑːsk/ *n.* 面具
electrical shock 电击	muscle strain 肌肉拉伤
relate to 与……相关	protective clothing 防护服
construction trade 建筑行业	

Notes. 注释。

1. They may work in narrow spaces or awkward positions. 他们可能在狭小的空间或令人尴尬的环境中工作。
2. Because most of their work is in buildings, installers and repairers lose less work

time because of bad weather than most other workers do in the construction trades. 因为他们的大部分工作在室内,比起建筑行业大多数其他工人,电梯安装员和维修工由于天气原因延误的时间会更少。

Answer the following questions. 回答以下问题。

1. What spaces and positions may elevator installers work?
_____

2. What dangers may they face?
_____

3. What do elevator workers wear when they work?
_____

4. Does the work of elevator installers belong to construction trade?
_____

3B: Read the passage and finish the exercise. 阅读文章,完成练习。

<h3 style="text-align:center">The Machine Room-Less (MRL) Installation</h3>

MRL elevator—as its name, it doesn't need a machine room. The new elevator system saves space, is far more energy efficient and avoids pollution. The costs of MRL installation are less than those traditional elevators.

MRL installations require fewer materials and less work time; No well holes to be drilled; no pits to be waterproofed. Some MRL installations don't need to hoist machines or control equipment to the penthouse floor.

Installation for MRL technology is highly visible and therefore offers more control over the work environment.

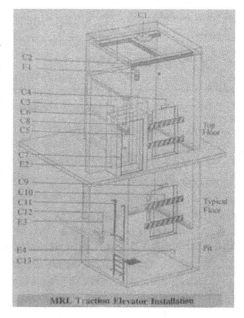

Word Bank. 词汇表。

> installation /ˌɪnstəˈleɪʃn/ *n.* 安装；设置　　drill /drɪl/ *v.* 钻孔
> visible /ˈvɪzəbl/ *adj.* 看得见的　　energy efficient 节约能源

Notes. 注释。

1. MRL elevator—as its name, it doesn't need a machine room. The new elevator system saves space, is far more energy efficient and avoids pollution. 无机房电梯,正如它的名字,是不需要机房的。新的电梯系统节约了空间,节约了更多的能源,还避免了污染。
2. Some MRL installations don't need to hoist machines or control equipment to the penthouse floor. 有的无机房安装不需要曳引机或楼顶机房的控制设备。
3. Installation for MRL technology is highly visible and therefore offers more control over the work environment. 无机房安装技术是显而易见的,所以也为工作环境提供更多的可控性。

Decide the sentences true (T) or false (F). 判断句子正误。

1. MRL installation needs less machine room.　　　　　　　　(　　)
2. MRL installation produces more pollution.　　　　　　　　(　　)
3. MRL installation requires no materials and work time.　　(　　)
4. Now all elevators are MRL elevators.　　　　　　　　　　(　　)
5. MRL technology gives more control over the work environment.　(　　)

Section D　Elevator Culture

4A：Magic Elevators. 神奇的电梯。

Taipei 101 Elevator in Taiwan

The world's fastest elevator is installed at Taiwan's new Taipei 101 tower. The Taipei 101 is 1,667 feet or 509 meters, a 101-storey building and has 67 elevator units, including two that service the 89th-floor observation deck and qualify as the world's fastest. The elevator system pushes the limits of people-mover technology. These rockets skyward at a peak speed of 3,314 ft. per minute (fpm) or 1,010 meters per minute or 60 km/hour, more than 800 fpm faster than the previous record holder in Japan's Yokohama Landmark Tower. By comparison, an airline pilot normally maintains a climb, or descent rate, of no more than 1,000 fpm.

Word Bank. 词汇表。

storey /ˈstɔːri/ n. 楼层　　　　　　　　descent /dɪˈsent/ n. 下降
push the limit 推向极限　　　　　　　peak speed 巅峰速度

Notes. 注释。

1. The Taipei 101 is 1,667 feet or 509 meters, a 101-storey building and has 67 elevator units, including two that service the 89th-floor observation deck and qualify as the world's fastest. 台北101大楼有1667尺或509米高，是101层的建筑物，有67个电梯单元，包括第89层的观景台。它被认为是世界上最快的电梯。

2. The elevator system pushes the limits of people-mover technology. 该电梯系统将人类的移动技术推至极限。

3. These rockets skyward at a peak speed of 3,314 ft. per minute (fpm) or 1,010 meters per minute or 60 km/hour, more than 800 fpm faster than the previous record holder in Japan's Yokohama Landmark Tower. 这些"火箭"以每分钟3314尺或每分钟1010米或每小时60千米的速度朝着天空呼啸而去,比前世界纪录保持者日本横滨的地标建筑塔电梯快每分钟800尺。

4. By comparison, an airline pilot normally maintains a climb, or descent rate, of no more than 1000 fpm. 相比之下,飞行员一般保持飞机爬升或下降的速度是每分钟1000尺。

Translation. 参考译文。

台湾台北101电梯

世界上最快的电梯是被安装在台湾的新台北101大楼。台北101大楼有1667尺或509米高,是101层的建筑物,有67个电梯单元,包括第89层的观景台。它被认为是世界上最快的电梯。该电梯系统将人类的移动技术推至极限。这些"火箭"以每分钟3314尺或每分钟1010米或每小时60千米的速度朝着天空呼啸而去,比前世界纪录保持者日本横滨的地标建筑塔电梯快每分钟800尺。相比之下,飞行员一般保持飞机爬升或下降的速度是每分钟1000尺。

4B: More About Elevator. 电梯知识。

十大电梯品牌
Ten Top Elevator Companies

No.1

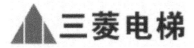

三菱电梯(上海三菱电梯有限公司)Shanghai Mitsubishi Elevator Co., Ltd. (SMEC)

成立于1987年,由上海机电股份有限公司与日本三菱电机等四方合资成立,是国内规模较大的电梯制造销售企业。该公司产品市场占有率自1941年到今在国内市场保持领先地位,是全国最大的500家外商投资企业之一。

No.2

OTIS 奥的斯(奥的斯电梯(中国)投资有限公司)Otis Elevator Company,Ltd.(China)

始于1853年,美国,电梯行业全球著名品牌。奥的斯电梯公司是全球专业的电梯、自动扶梯和自动人行道的制造商和服务提供商,专注服务于建筑领域。1853年,公司创始人奥的斯先生发明了安全电梯。目前奥的斯的业务遍及全球超过200个国家和地区,并为全球超过190万台电梯和自动扶梯提供维护保养服务。

No.3

通力KONE(通力电梯有限公司)Giant KONE Elevator Co.,Ltd.

始创于1910年,芬兰,电梯十大品牌,全球电梯和自动扶梯产业大型供应商,其无齿轮电梯在业界享有盛名。巨人通力电梯有限公司是由芬兰通力集团和中国浙江巨人电梯有限公司于2005年在中国投资兴建的合资企业。成立于1910年的芬兰通力集团是世界最大的电梯和自动扶梯公司之一,于1996年进入中国市场并以率先发明的小机房和无机房等革命性技术为其赢得了后发优势,获得了国家大剧院、首都国际机场、国家体育场等一系列经典工程项目。

No.4

Hitachi 日立(日立(中国)有限公司)HITACHI Elevator Co.,Ltd.

始于1910年,日本,世界500强,全球著名的日本工业品牌的代表,日本较大的综合电机生产商。日立以过百年的研发技术进入中国,投资建立了日立海外最大的电梯公司——日立电梯(中国)有限公司。日立电梯多年来一直致力于各类电梯、扶梯、自动人行道等的研发、制造、销售、安装、维修、保养以及进出口贸易服务,超过 64 个营分司遍布全国各个主要城市,并在广州、天津、上海、成都四个地方分别成立了大型制造基地。目前年产能超过 6 万台,是国内最大的电梯制造商和服务商之一。

No.5

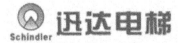

Schindler 迅达(迅达(中国)电梯有限公司)Schindler Elevator Co., Ltd. (China)

创立于 1874 年,瑞士,全球知名的电梯、自动扶梯及相关服务的供应商,电梯十大品牌。迅达于 1980 年进入中国,即迅达(中国)电梯有限公司。在服务中国的 30 多年里,迅达(中国)持续不懈地满足广大客户群体的需求,以优质的产品和服务为中国的高层地标、商业地产、公共交通等建筑做出了贡献。同时迅达也致力于对中国的员工、服务、制造和研发等各方面的进一步投资。2014 年 5 月,位于上海嘉定的全球扶梯新工厂正式投产,它见证了瑞士迅达进一步扎根中国、立足长远的意愿。

No.6

蒂森克虏伯(蒂森克虏伯电梯(上海)有限公司)Thyssenkrupp Elevator Technology (Shanghai Co,. Ltd.)

始于 1865 年,德国,电梯十大品牌,蒂森克虏伯集团旗下,全球领先的电梯和自动扶梯生产商。蒂森克虏伯电梯 1995 年进入中国,随着业务和生产能力的快速增长,我们现在全国有约 10,000 名员工,四个生产基地,公司产品包括客用及货用电梯、自动扶梯、自动人行步道、旅客登机桥、座椅电梯及升降平台,并为各种产品提供量身定制的服务方案。依靠密集的分公司、办事处网络,技术雄厚的本地化工厂和全球技术服务中心,蒂森克虏伯电梯(上海)为所有的产品系列提供新梯安装、更新改造和维修保养服务,随时随地,无处不在。

No.7

西子奥的斯 XiziOTIS（奥的斯电梯（中国）投资有限公司）XiziOtis Elevator Company, Ltd. (China)

全球较大的电梯制造商奥的斯旗下，知名垂直升降电梯、自动扶梯、自动人行道制造商和服务提供商。西子奥的斯电梯有限公司是世界第一电梯品牌奥的斯在中国最大的控股子公司，是中国绿色电梯第一品牌和中国最大的电扶梯制造商及服务商之一，是世界500强企业，是联合技术公司建筑与工业系统旗下奥的斯电梯的在华子公司。西子奥的斯以敏锐的洞察力和专业创新的精神，致力于提供最值得信赖的、对环境最友好的建筑移动解决方案，持续为顾客创造最优价值，成就员工自我发展，主动承担社会责任，并以良好的回报坚定投资者的信心。"以客为先，服务至上"是西子奥的斯一贯坚持的服务理念。西子奥的斯坚持以客户为导向，对服务品质孜孜以求，为客户提供"邻居式"的金牌服务。

No.8

富士达 FUJITEC（华升富士达电梯有限公司）HUASHENG FUJITEC ELEVATOR CO.,LTD.

始建于1948年，日本，电梯十大品牌，专业从事电梯、自动扶梯、自动人行道、立体停车设备等空间移动系统的全球性跨国企业。华升富士达电梯有限公司是把电梯专业生产厂家富士达公司多年培植起来的经营能力和技术资本与中国中纺集团公司的销售网络、多元化战略体系有机地结合起来，集"科、工、贸"于一体的高科技型企业。多年来富士达向中国各地销售了许多用于高层建筑的高速电梯，其中具代表性的有"中国国际贸易中心""北京贵宾楼饭店""上海证券大厦""上海商城"等。在这些最具现代化的高楼之中活跃着为数众多的"富士达"高级电梯和自动扶梯。

富士达坚持以人为本,以技术为本,以产品为本,适应新时代,在世界各国,与世界人民一起,创造卓越的城市功能。

No.9

TOSHIBA 东芝电梯

Toshiba 东芝电梯(东芝(中国)有限公司)Toshiba Elevator Co., Ltd.

创立于1875年,日本,世界电梯产业的著名企业,集升降机相关产品系统开发、制造、安装、调试、维修服务于一体。2001年,新的"东芝电梯株式会社"成立(简称TELC)。东芝(中国)有限公司成立于1995年,主要从事"TOSHIBA"品牌电梯(扶梯)的开发、设计、销售、制造、安装、维保及改造等业务。作为跨国企业,公司致力于开发制造高水准的环境友好型电梯产品。东芝(中国)有限公司是东芝电梯集团公司开展研发高尖端电梯技术的核心基地,东芝电梯的心脏机构—东芝电梯全球研发中心就设在公司内。公司先后取得了国际认证机构—英国劳氏LRQA颁发的ISO9001国际质量管理体系认证,ISO4001国际环境管理体系认证以及OHSAS18001职业健康安全管理体系的认证。

No.10

康力Canny(康力电梯股份有限公司)Canny Elevator Co., Ltd.

成立于1997年,行业标准参编单位,集电梯设计、开发、制造、销售、安装和维保于一体的高新技术企业。康力电梯创建的富有特色的企业文化,是在充分吸收其他先进企业文化的基础上,结合自身特点,进行优化、凝练和设计,而形成的一套对于企业意识和行为极具实际指导意义的理念体系。社会责任是康力电梯的第一责任。康力电梯不断提高自身效益,为地方经济和行业发展做出卓越贡献,大大提升民族电梯品牌的社会地位。

4C: Evaluation. 回顾所学知识，写下你已掌握的单词和句子。

Write down the words that you think are the most useful:

1. _____
2. _____
3. _____
4. _____
5. _____
6. _____
7. _____
8. _____

List the sentences that you think are the most important:

1. _____
2. _____
3. _____
4. _____
5. _____
6. _____
7. _____
8. _____

Chapter 5　　Elevator Test and Inspection

Section A Warming Up

1A: Match and say. 根据所给单词和图片,配对并说出相应的名称。

| clamp | landing | overload |
| full load | no load | handrail |

_____

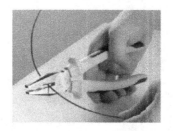
_____

_____

_____

_____

_____

| Chapter 5 | Elevator Test and Inspection

1B: Guess and read. 识别电梯标识并朗读。

| examine and test | spark | deliver |
| engage | strike | clamp |

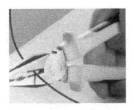

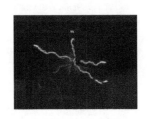

1C: Match and say. 连线并朗读。

1. specification   A. 接地
2. audible   B. 额定负载
3. leveling accuracy   C. 锁紧元件
4. rated load   D. 说明书
5. earthing   E. 验收
6. locking element   F. 平层精度
7. accept   G. 听得见的

1D：Word Bank. 词汇表。

handrail /ˈhændreɪl/ n. 扶手	clamp /klæmp/ n. 夹钳；v. 夹住
buffer /ˈbʌfə/ n. 缓冲器	governor /ˈgʌvənə/ n. 限速器
landing /ˈlændɪŋ/ n. 层站；楼梯平台	spark /spɑːk/ v. 产生火花；n. 火花
deliver /dɪˈlɪvə/ v. 输送	accept /əkˈsept/ v. 验收
engage /ɪnˈgeɪdʒ/ v. 咬合，(使)衔接	strike /straɪk/ v. 碰；撞击
specification /ˌspesɪfɪˈkeɪʃn/ n. 说明书	audible /ˈɔːdəbl/ adj. 听得见的
earthing /ˈɜːθɪŋ/ n. 接地	examine and test 检测
safety gear 安全钳	guide rail 导轨
leveling accuracy 平层精度	rated load 额定负载
locking element 锁紧元件	

Section B Useful Sentences

2A: Match and write sentences. 根据图片,选写合适的句子。

> A. All equipment and systems shall be examined and tested after installation work.
> B. Elevator should also be given a 60-minute test with full load. It should stop at all landings opening and closing, up and down.
> C. Generally, elevator equipment shall run under all loading conditions from no load to 10% overload in order to check its operation and level accuracy.
> D. The car shall not be able to start until the locking elements are engaged by at least 7mm.
> E. During the test, the equipment should not overheat, spark excessively, become noisy.
> F. If the car is overloaded, the computer will not close the doors until some of the weight is removed.

1.  _____

2.  _____

3.  _____

4.

5.

6.

2B: Translate the following sentences. 翻译下列句子，注意画线部分词语。

1. Please follow the specification to install and use the elevator.

2. Elevator alerts should be heard with an audible sound.

3. The check and accept of the elevator includes the check and accept of the installation and function.

4. A passenger and freight elevator is a lift designed to deliver both people and goods.

5. Electrical devices include earthing, safety devices and so on.

2C: Word Bank. 词汇表。

equipment /ɪˈkwɪpmənt/ n. 装备；设备   run /rʌn/ v. 运行
fully load 满员   overload 超载

Section C Passages

3A: Read the passage. 阅读文章,回答问题。

Elevator Test Details

Before an elevator is put into service, some examinations and tests should be carried out, such as balance test, full load test, speed test, temperature rise test, car leveling test, brake test, locking test and so on.

For the car leveling test, regardless of load in car or direction of travel, floor leveling accuracy of any landing floor should be within plus or minus 3 mm (about 1/8 inch).

The brake system examination and test should be carried out when the car is going down at rated speed with 125% of the rated load and interrupting the supply to the motor and the brake.

For locking devices, the car should not be able to start until the locking elements are engaged by at least 7mm.

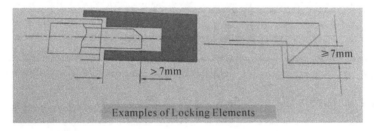

Examples of Locking Elements

Word Bank. 词汇表。

detail /ˈdiːteɪl/ n. 细节	interrupt /ˌɪntəˈrʌpt/ v. 使中断;插嘴
plus /plʌs/ adj. 零上;好的; prep. 加	minus /ˈmaɪnəs/ adj. 负的;小于零的
put into service 投入使用	carry out 执行
temperature rise test 升温测试	regardless of 无视,不管

Notes. 注释。

1. For the car leveling test, regardless of load in car or direction of travel, floor leveling accuracy of any landing floor should be within plus or minus 3 mm (about 1/8 inch). 就轿厢水平检测而言，不管轿厢的运载量或运行方向，任何楼层的平层精度都应该在正负 3 毫米或正负八分之一英寸。

2. The brake system examination and test should be carried out when the car is going down at rated speed with 125％ of the rated load and interrupting the supply to the motor and the brake. 制动器系统的检查和测试应该在以下情况执行：轿厢以 125％ 的额定载容量在额定速度下行时切断电马达和制动器。

Answer the following questions. 回答下列问题。

1. What examinations and tests should be carried out before an elevator is put into service?

_____

2. What do the numbers 3，125 and 7 mean in the passage?

Number	Meaning
3	
125	
7	

3B: Read and finish the exercise. 阅读文章，完成练习。

Otis plans tallest elevator test tower

Elevator giant Otis is to build the world's tallest elevator test tower in Shanghai, as it works to keep pace with China's fast developing skyscrapers. The tower will be part of the company's new global research and development facilities in the city, which will be finished by the end of 2018 with its height of around 270 meters.

"As the industry leader, we introduce safe-elevator technology. This is a strong example of our new idea and technology development for our next generation of elevators and service." said Otis President Philippe Delpech.

Elevator test tower is a structure usually 100 to 180 meters tall, which is designed to test the stress and limits of special elevator cars. Tests are also carried out in the tower to insure reliability and safety in elevator designs and solve any problems.

Word Bank. 词汇表。

giant /ˈdʒaɪənt/ n. 巨人
facility /fəˈsɪləti/ n. 设施；设备
industry /ˈɪndəstri/ n. 工业；行业
stress /stres/ n. 压力
reliability /rɪˌlaɪəˈbɪləti/ n. 可靠性
keep pace with 跟上步伐
Philippe Delpech 菲利普戴培杰（奥的斯电梯全球总裁）

Otis 奥的斯电梯公司
global /ˈgləʊbl/ adj. 全球的
structure /ˈstrʌktʃə/ n. 结构
limit /ˈlɪmɪt/ n. 限度
elevator test tower 电梯测试塔

Notes. 注释。

1. Elevator giant Otis is to build the world's tallest elevator test tower in Shanghai, as it works to keep pace with China's fast developing skyscrapers. 为了跟上中国高速发展高层建筑的步伐，电梯巨人奥的斯公司将在中国上海建造世界最高的电梯检测塔。

2. The tower will be part of the company's new global research and development facilities in the city, which will be finished by the end of 2018 with its height of around 270 meters. 作为奥的斯公司新的全球城市研究和发展设施的一部分，该电梯检测塔将于 2018 年年底竣工，高度大约为 270 米。

3. This is a strong example of our new idea and technology development for our next generation of elevators and service. 这是我们公司新一代电梯和服务新理念和技术发展的一个有力证明。

4. Tests are also carried out in the tower to insure reliability and safety in elevator

designs and solve any problems. 检测塔还负责检测电梯设计的可靠性、安全性以及解决可能出现的任何问题。

Answer the following questions. 回答以下问题。

1. Where will the world's tallest elevator test tower be?
   _____

2. When will the elevator test tower be finished?
   _____

3. What will its height be?
   _____

4. Who is Philippe Delpech?
   _____

5. What is an elevator test tower designed to?
   _____

Section D Elevator Culture

4A：Magic Elevators. 神奇的电梯。

Santa Justa Lift in Portugal

The Santa Justa Lift (Elevador de Santa Justa or do Carmo), is a lift in the city of Lisbon at Santa Justa Street. It connects downtown streets with the uphill Carmo Square.

The Santa Justa Lift was designed by Raul Mesnier de Ponsard. The construction began in 1900 and was finished in 1902. Originally powered by steam, it was powered by electrical operation in 1907.

The iron lift is 45 metres tall and is designed in neogothic style with a different pattern on each storey. The top storey is reached by helicoidal staircases and has a flat that gives views of Lisbon Castle, the Rossio Square and the Baixa neighbourhood. There are two elevator booths. Each booth has a wooden interior for 20 people. The lift has become a tourist attraction in Lisbon, and among the urban lifts in Lisbon, Santa Justa is the only vertical one.

Word Bank. 词汇表。

originally /əˈrɪdʒənəli/ *adj*. 起初
booth /buːð; NAmE buːθ/ *n*. 不受干扰的划定空间（如电话亭、投票间等）
interior /ɪnˈtɪəriə/ *adj*. 在内的；内部的   urban /ˈɜːbən/ *adj*. 城市的
helicoidal /ˌhelɪˈkɔɪdəl/ *adj*. 螺旋状的   vertical /ˈvɜːtɪkl/ *adj*. 垂直的
Lisbon /ˈlizbən/ 里斯本（葡萄牙首都）   uphill /ˌʌpˈhɪl/ *adj*. 上坡的
Carmo Square 卡莫广场   electrical operation 电气操作
neogothic style 新哥特式风格   Santa Justa Lift 圣胡斯塔电梯
Lisbon Castle 里斯本城堡   the Rossio Square 罗西乌广场

Notes. 注释。

1. The Santa Justa Lift (Elevador de Santa Justa or do Carmo), is a lift in the city of Lisbon at Santa Justa Street. It connects downtown streets with the uphill Carmo Square. 圣胡斯塔电梯(葡萄牙语：Elevador de Santa Justa 或者说 do Carmo)是葡萄牙首都里斯本圣胡斯塔街的一座电梯。它连接着上坡山的卡莫广场和市中心的街道。

2. Originally powered by steam, it was powered by electrical operation in 1907. 刚开始的时候，电梯使用蒸汽动力，于 1907 年改为使用电力。

3. The iron lift is 45 metres tall and is designed in neogothic style with a different pattern on each storey. 这台钢铁电梯高 45 米，新哥特式风格，每层不同的样式。

4. The top storey is reached by helicoidal staircases and has a flat that gives views of Lisbon Castle, the Rossio Square and the Baixa neighbourhood. 通过螺旋楼梯可以到达顶层，有一个阳台，可供观赏里斯本城堡、罗西乌广场和庞巴尔下城的景色。

5. There are two elevator booths. Each booth has a wooden interior for 20 people. 电梯设有两个升降机笼，均为木质内饰，最多可乘载 20 名乘客。

6. The lift has become a tourist attraction in Lisbon, and among the urban lifts in Lisbon, Santa Justa is the only vertical one. 这台电梯已经成为里斯本的一大旅游景观，在里斯本的城市电梯中，圣胡斯塔是唯一的一台垂直电梯。

Translation. 参考译文。

葡萄牙圣胡斯塔电梯

圣胡斯塔电梯(葡萄牙语：Elevador de Santa Justa 或者说 do Carmo)是葡萄牙首都里斯本圣胡斯塔街的一座电梯。它连接着上坡山的卡莫广场和市中心的街道。

圣胡斯塔电梯的设计者是 Raul Mesnier de Ponsard。电梯始建于 1900 年，完成于 1902 年。刚开始的时候，电梯使用蒸汽动力，于 1907 年改为使用电力。

这台钢铁电梯高 45 米，新哥特式风格，每层不同的样式。通过螺旋楼梯可以到达顶层，有一个阳台，可供观赏里斯本城堡、罗西乌广场和庞巴尔下城的景色。电梯设有两个升降机笼，均为木质内饰，最多可乘载 20 名乘客。这台电梯已经成为里斯本的一大旅游景观，在里斯本的城市电梯中，圣胡斯塔是唯一的一台垂直电梯。

4B: More About Elevator. 电梯知识。

十大电梯品牌
Ten Top Elevator Companies

三菱电梯的正式公司总部在日本,公司建立于 1973 年。

三菱电梯公司 Mitsubishi Elevator

三菱电梯原装梯在中国有两家厂商代理,分别为广东菱电电梯有限公司与上海三菱电梯有限公司。

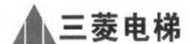　　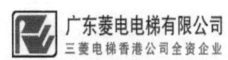

上海三菱电梯有限公司　　　　　　　　广东菱电电梯有限公司
Shanghai Mitsubishi Elevator Co., Ltd. (SMEC)　　Guangdong rhombus Electric Elevator Co., Ltd.

广东菱电电梯有限公司(广东菱电)成立于 1998 年,为三菱电梯香港有限公司在中国内地的全资企业,是中国内地生产三菱牌电梯的企业。公司总部位于广东省中山市,在北京、上海、江苏、浙江、大连、青岛、哈尔滨、广州、深圳、成都设有分公司,业务遍及中国内地各省市及港澳地区。专业经营日本三菱电机株式会社生产的三菱牌电梯及自动扶梯,并提供立体停车库、垂直及横向运输系统等产品的生产、加工、组装、营销、安装、维修及保养一条龙服务。

广东菱电电梯有限公司

1987年,由上海机电股份有限公司与日本三菱电机等四方合资成立上海三菱电梯有限公司。它是国内规模较大的电梯制造销售企业。该公司产品市场占有率自1941年到今在国内市场保持领先地位,是全国最大的500家外商投资企业之一。

上海三菱电梯有限公司

　　上海三菱电梯有限公司的前身是上海航空工业中等技术学校、上海市劳动第一技工学校。在1969年两校合并建厂,成立长城机械厂。1981年改名为上海长城电梯厂,成为国内生产电梯的骨干企业之一。公司经过几十年的创业与发展,已成为中国最大的电梯制造和销售企业之一。公司产品市场占有率已连续多年在中国电梯市场保持领先地位。

　　追求卓越的企业理念是企业发展战略的源头。在企业发展过程中,面对不断变化的市场环境,从用户的需要就是上海三菱的追求到以顾客为中心向创造更和谐的生活空间的企业理念的发展,为企业持续快速的发展提供了根本保证。未来公司的理念将不断向国际化经营转变,并积极探索符合时代发展需要的企业理念。

　　合资后相当一段时期内的企业精神:"精诚团结、努力拼搏"。进入21世纪的企业精神:"团结、敬业、自律、创新"。自公司成立以来,在各方携手合力和企业技术中心(国家级)的集体努力下,公司加快引进和转化日本三菱具有世界领先水平的全电脑交流变压变频(VVVF)电梯技术,成为国内首家推出VVVF系列电梯制造企业。在此基础上,公司持续动态引进和转化世界领先的最新技术全电脑控制智能化系列电梯和新型的自动扶梯系列,并与日本三菱电机联手开发优质的住宅电梯。近年来,公司又加快转化了ELENSSA最新无机房和NEXWAY小机房等具有节能、环保的电梯技术。

　　在加快引进、转化的同时,公司加大自主开发的力度,先后开发了先进技术水平的微机网络控制交流变压变频 HOPE 系列和菱云(LEHY)系列电梯,自主知识产权产品占销售总量的比例逐年提高。2006 年自主知识产权产品占销售总量的 60%。20 多年来,公司以技术创新为核心,坚持做到引进与开发并重,使电梯技术始终保持国内领先、国际先进水平。目前,公司已发展到 34 个产品系列、200 多种不同规格的产品,形成了覆盖市场不同层次需求的可供产品体系。

20 年来,公司坚持以顾客为中心,贯彻"全面覆盖、纵深发展"和"精耕细作"的营销理念,不断建立和完善市场营销运作体制和服务体系,在全国成立了 7 大区域和 36 个直属分公司,并建立了 230 余个销售、安装和维修服务代理点。

公司不断吸收和借鉴世界上先进的经营理念、管理方法,努力实现企业管理与国际接轨。公司先后通过 ISO9001、ISO14001 和 OHSAS18001 三个管理体系的认证以及欧盟 CE 电梯指令认证。

公司加快推进企业信息化建设,实施 CIMS 应用工程,经过全面规划,循序渐进,已建立覆盖公司本部和全国分公司信息化网络。这个集成、同步、高效、稳定的信息化网络,全天候以先进的技术和科学的管理理念支撑着企业的经营和运作。

以下是三菱电梯公司的典型项目:

1. 上海金茂大厦

金茂大厦位于上海浦东陆家嘴,是一座 88 层的超高层大厦,建筑高度 420.5 平方米,建筑面积 28.9 万平方米,是目前亚洲第三、中国第一高楼。它是中国改革开放、经济腾飞的象征之一。大厦全部选用三菱垂直电梯及自动扶梯,共计 79 台,其中 2 台为目前中国大陆最高速的 9m/s 电梯。

上海金茂大厦

2. 上海世博中心

上海世博会是国家项目,世博中心位于世博园区浦东核心地块 B 区沿江位置。

作为世博会永久性场馆中重要场馆之一的世博中心以会议接待、公共活动为主。世博会期间，它将作为园区的庆典中心、文化交流中心、新闻中心、接待宴请中心和指挥运营中心。世博会以后将作为高规格国际性和国内重要论坛和会议的场所，提供一流的国际会议配套设施。该项目全部选用了三菱电梯和三菱自动扶梯，共 65 台，并已委托上海三菱电梯有限公司负责安装及承担长期的维护保养工作。

上海世博中心

3. 深圳卓越皇岗世纪中心

卓越皇岗世纪中心位于深圳 CBD 南区东南部，占地 3 万平方米，总建筑面积 43 万平方米，其主楼高 300m，集写字楼、商务公寓、商业、酒店多种业态为一体，为深圳福田中心区迄今最大的都市综合体项目。卓越皇岗世纪中心全部选用了三菱电梯和三菱自动扶梯，共 116 台，其中 9 台为 6m/s 电梯。该项目已委托上海三菱电梯有限公司负责安装及承担长期的维护保养工作。

深圳卓越皇岗世纪中心

上海三菱战略目标是成为"国际区域性知名公司",尽快进入世界知名电梯企业行列。为实现这一目标,上海三菱电梯有限公司将继续发扬"团结、敬业、自律、创新"精神,实施"四个战略",坚持"超越自我,从零开始"为核心的企业文化,为发展中国电梯工业多做贡献。

4C:Evaluation. 回顾所学知识,写下你已掌握的单词和句子。

Write down the words that you think are the most useful:

1. _____
2. _____
3. _____
4. _____
5. _____
6. _____
7. _____
8. _____

List the sentences that you think are the most important:

1. _____
2. _____
3. _____
4. _____
5. _____
6. _____
7. _____
8. _____

Chapter 6　Troubleshooting and Maintenance

| Chapter 6 | Troubleshooting and Maintenance

Section A　Warming Up

1A：Match and say. 根据所给单词和图片，配对并说出相应的名称。

security contact	fuse
photo eye	bearing
lead	rotary encoder
balance factor	traction sheave
short circuit	rating current

_____   _____

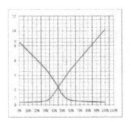

_____   _____

_____   _____

087

电梯英语 | *Elevator English For Vocational School Students*

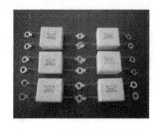

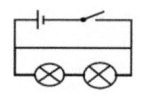

1B: Learn and read. 学习下列动词。

| jump | release | stall |
| blown | vibrate | coast |

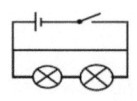

| Chapter 6 | Troubleshooting and Maintenance

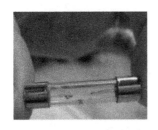

_____   _____

1C: Word Bank. 词汇表。

fuse /fjuːz/ v. (使)熔化；n. 保险丝	bearing /ˈbeərɪŋ/ n. 轴承
lead /liːd/ n. 电线	jump /dʒʌmp/ n. 短接
stall /stɔːl/ adj. 失控的	blown /bləʊn/ adj. 熔断的
coasting /ˈkəʊstɪŋ/ n. 溜梯	rotary encoder 旋转编码器
balance factor 平衡系数图	traction sheave 曳引轮
short circuit 短路	vibrate /vaɪˈbreɪt/ v. 震动
rating current 额定电流	security contact 安全触点
photo eye 光电检测器	

Section B Useful Sentences

2A: Match and write sentences. 根据图片,选写合适的句子。

> A. If there is any failure, elevator will emergency stop by any safety device and switch or contact in safety circuit.
>
> B. When closing, the door will reopen if safety edge or light curtain touches something.
>
> C. Occasionally elevator returns to the home landing, and the sudden coasting causes great fear to passengers.
>
> D. Operation failed usually because the user play safety devices and switch, or maintenance is not good at it and repairs blindly, such as door security contacts jump.
>
> E. We must try to avoid the elevator accident which there is an immediate threat to life.
>
> F. Make sure the building's main power supply is present, then check the meter room for any tripped breakers or blown fuses.

1. _____

2. _____

3. _____

| Chapter 6 | Troubleshooting and Maintenance

4.

5.

6.

2B: Translate the following sentences. 翻译下面的句子。

1. Exam that doors and gates are closed and nothing stops the movement of the car.

2. Don't move a stalled lift or remove a passenger when the car is not near the floor level.

3. When the button releases, lights don't light or don't stay lit.

4. For motor loads, a time delay fuse should have a rating of no less than 100% and no more than 125% of the full load running current.

5. Excessive car vibration is caused by either electrical (drive system) or mechanical problems in operation system.

2C: Word Bank. 词汇表。

troubleshooting /ˈtrʌblʃuːtɪŋ/ n. 故障
emergency /iˈmɜːdʒənsi/ n. 紧急情况
reopen /ˌriːˈəʊpən/ v. 再打开，重新打开
maintenance /ˈmeɪntənəns/ n. 保养；维护
blindly /ˈblaɪndli/ adv. 盲目地
excessive /ɪkˈsesɪv/ adj. 过分的
safety edge 安全触板

failure /ˈfeɪljə/ n. 失败
occasionally /əˈkeɪʒnəli/ adv. 偶然
verify /ˈverɪfaɪ/ v. 证实；查对
rating current 额定电流
tripped breakers 跳闸

Chapter 6 | Troubleshooting and Maintenance

Section C Passages

3A: **Practice the dialogue and answer the questions.** 练习对话，回答问题。

Jeff: Hi, Joan. You look a little nervous today.

Joan: I feel very anxious. The lift I was in this morning stopped between the 11th and the 12th floor!

Jeff: That sounds terrible. What did you do?

Joan: I followed the instructions in the lift and pushed the emergency button.

Jeff: And then what happened?

Joan: The technician told me to stay calm and relax, while they tried to repair it.

Jeff: How long did they take to get you out?

Joan: Only about 20 minutes, but I was very nervous.

Jeff: I would be too! Were you alone in the lift?

Joan: No. There was a guy in the lift with me. When the lift was fixed, he just pushed me aside and ran out of the elevator.

Jeff: What a rude guy!

Word Bank. 词汇表。

anxious /ˈæŋkʃəs/ *adj.* 忧虑的；焦急的   technician /tekˈnɪʃn/ *n.* 技术员
calm /kɑːm/ *adj.* 平静的，镇静的   push sb. aside 把某人推到一边

Notes. 注释。

1. I followed the instructions in the lift and pushed the emergency button. 我遵照电梯里的指示按下了紧急按钮。
2. The technician told me to stay calm and relax, while they tried to repair it. 技术人员告诉我保持并镇定放松，同时他们在努力维修电梯。

Answer the following questions. 回答下列问题。

1. What happened to Joan this morning?
   _____

2. Was she very nervous?
   _____

3. How did Joan do then?
   _____

4. What did the technician do?
   _____

5. How long did it take the technician to fix the lift?
   _____

6. What did Jeff think of the guy?
   _____

3B: Read the passage and finish the exercise. 阅读文章,完成练习。

Tips On Surviving In A Stuck Lift

Whether you call it a lift or an elevator, getting stuck in a lift is terrible. Here are some tips for you on surviving in a stuck lift.

Step 1: Remain calm

There are very few cases of people dying in stuck lifts. Modern lifts are always so safe to prevent them from plummeting even in the case of a power cut. It is very likely that the worst thing is that you'll have to wait a short while until the lift is working again. If you're in a lift that suddenly stops, try pushing a floor button again, or the open door button. If the lift still doesn't work, it's time to call for help.

Step 2: Call for help

There's a telephone or an alarm button to call for help. If you have a mobile phone, you can use it to call the emergency services or call a friend to ask him to do it for you.

You can also shout loudly or bang on the door to let someone hear outside, and wait a while patiently.

Step 3: Deal with loneliness

A long wait in a lift can be a very lonely time, and you may want someone to be with you. If this is the case, make sure that he doesn't panic. Try your best to keep him calm.

Thanks For Reading How To Survive In A Trapped Elevator.

Word Bank. 词汇表。

> stuck /stʌk/ *adj.* stick 的过去式(分词)，被卡住的；动不了的
>
> plummet /'plʌmɪt/ *v.* 垂直落下，骤然下跌
>
> patiently /'peɪʃntli/ *adv.* 耐心地   panic /'pænɪk/ *adj.* 惊恐
>
> trap /træp/ *v.* 使陷入绝境   alarm button 警示按钮
>
> emergency services 紧急服务

Notes. 注释。

1. There are very few cases of people dying in stuck lifts. 几乎没有什么因被困电梯而死亡的案例。

2. It is very likely that the worst thing is that you'll have to wait a short while until the lift is working again. 最糟糕的情况是你不得不在电梯中待上一小会直到电梯重新恢复运行。

3. If you're in a lift that suddenly stops, try pushing a floor button again, or the open door button. 如果你乘坐的电梯忽然停止运行了，试着再次按楼层按钮或者按开门按钮。

4. You can also shout loudly or bang on the door to let someone hear outside, and wait a while patiently. 你也可以高声叫喊或梆梆敲门，好让外面的人听到，然后耐心等待。

5. If this is the case, make sure that he doesn't panic. Try your best to keep him calm. 在这种情况下，确保对方不要惊恐，尽力让他保持镇定。

Answer the following questions. 回答下列问题。

1. What are the 3 tips for you on surviving in a stuck lift?
   _____

2. Are there many cases of people dying in stuck lifts?
   _____

3. What is likely the worst thing when you are in a stuck lift?
   _____

4. If you're in a lift that suddenly stops, what should you do first?
   _____

5. What can you do to call for help?
   _____

6. How do you deal with loneliness?
   _____

Section D　Elevator Culture

4A：Magic Elevators. 神奇的电梯。

Hammetschwand Elevator in Switzerland

Europe's highest exterior elevator, the Hammetschwand Lift is in Switzerland. It connects a very large rock path with the lookout point Hammetschwand on the Burgenstock plateau overlooking Lake Lucerne. It brings passengers 153 meters up to the summit of the Hammetschwand in less than one minute. At its time it had a speed of one meter per second and one could enjoy nearly three minutes of travel. Its wood cab could carry 8 passengers. During the upgrade of 1935 the speed was 2.7 meters per second and the cab was replaced with a light metal. It was not only the highest public external elevator of Europe, but also the fastest elevator of the world.

Word Bank. 词汇表。

Burgenstock 比尔根山	Switzerland /ˈswɪtsələnd/ n. 瑞士
exterior /ɪkˈstɪərɪə/ adj. 外部的	plateau /ˈplætəʊ/ n. 高原
overlook /ˌəʊvəˈlʊk/ v. 俯视	summit /ˈsʌmɪt/ n. 峰会

> upgrade /ˌʌpˈgreɪd/ v. 升级
> Hammetschwand Elevator 哈梅茨施万德观光电梯
> Lake Lucerne 琉森湖

Notes. 注释。

1. It connects a very large rock path with the lookout point Hammetschwand on the Burgenstock plateau overlooking Lake Lucerne. 一条巨大的岩石路连着哈梅茨施万德观光台，登上高处，你可以尽情浏览比尔根山高原上的琉森湖。

2. At its time it had a speed of one meter per second and one could enjoy nearly three minutes of travel. 以前电梯速度是每秒钟一米，你可以享受将近 3 分钟的过程。

3. During the upgrade of 1935 the speed was 2.7 meters per second and the cab was replaced with a light metal. 1935 年，电梯速度得到提升，达到每秒 2.7 米，机笼也替换为轻金属材质。

Translation. 参考译文。

瑞士哈梅茨施万德观光电梯

欧洲最高的户外电梯是瑞士的哈梅茨施万德观光电梯。一条巨大的岩石路连着哈梅茨施万德观光台，登上高处，你可以尽情浏览比尔根山高原上的琉森湖。电梯带着哈梅茨施万德峰会的游客在 1 分钟内登上高达 153 米的高处。以前电梯速度是每秒钟一米，你可以享受将近 3 分钟的过程。它的机笼是木质的，可以搭乘 8 名乘客。1935 年，电梯速度得到提升，达到每秒 2.7 米，机笼也替换为轻金属材质。它不仅是欧洲最高的公共户外电梯，也是世界上运行速度最快的电梯。

4B：More About Elevator. 电梯知识。

电梯常用术语（中英文）

1. 电梯 LIFT；ELEVATOR

服务于规定楼层的固定式升降设备。它有一个轿厢，运行在至少两列垂直的或倾斜角小于 15 度的刚性轨之间。轿厢尺寸与结构形式便于乘客出入或装货物。

2. 乘客电梯 PASSENGER LIFT

为运送乘客而设计的电梯。

3. 载货电梯 GOODS LIFT; FREIGHT LIFT

通常有人伴随，主要为运送货物而设计的电梯。

4. 客货电梯 PASSENGER-GOODS LIFT

以运送乘客为主，但也可运送货物的电梯。

5. 病床电梯；医用电梯 BED LIFT

为运送病床（包括病人）及医疗设备而设计的电梯。

6. 住宅电梯 RESIDENTIAL LIFT

供住宅楼使用的电梯。

7. 杂物电梯 DUMBWAITER LIFT; SERVICE LIFT

服务于规定楼层的固定式升降设备。它有一个轿车厢，就其尺寸和结构形式而言，轿厢内不允许进人。轿厢运行在两列垂直的或倾斜角15度的刚性导轨之间。为满足不得进入的条件，轿厢尺寸不得超过：

A) 底板面积：1m×1m

B) 深度：1m

C) 高度：1.2m

但是，如果轿车厢由几个永久的间隔组成，而每一个间隔都能满足上述要求，则高度超过1.2m是允许的。

8. 船用电梯 LIFT ON SHIPS

船舶上使用的电梯。

9. 观光电梯 PANORAMIC LIFT; OBSERVATION LIFT

井道和轿厢壁至少有同一侧透明，乘客可观看轿车厢外景物的电梯。

10. 汽车电梯 MOTOR VEHICLE LIFT；AUTOMOBILE LIFT

为运送车辆而设计的电梯。

11. 液压电梯 HYDRAULIC LIFT

依靠液压驱动的电梯。

12. 平层精度 LEVELING ACCURACY

轿厢到站停靠后，轿厢地坎上平面与层门地坎上平面之间垂直方向的偏差值。

13. 电梯额定速度 RATED SPEED OF LIFT

电梯设计所规定的轿车厢速度。

14. 检修速度 INSPECTION SPEED

电梯检修运行时的速度。

15. 额定负载 RATED LOAD；RATED CAPACITY

电梯设计所规定的轿厢内最大载荷。

16. 电梯提升高度 TRAVELING HEIGHT OF LIFT；LIFTING HEIGHT OF LIFT

从底层端站楼面至顶层端站楼面之间的垂直距离。

17. 机房 MACHINE

安装一台或多台曳引机其及附属设备的专用房间。

18. 机房高度 MACHINE ROOM HEIGHT

机房地面至机房顶板之间的最小垂直距离。

19. 机房宽度 MACHINE ROOM WIDTH

机房内沿平行于轿厢宽度方向的水平距离。

20. 机房深度 MACHINE ROOM DEPTH

机房内垂直于机房宽度的水平距离。

21. 机房面积 MACHINE ROOM AREA

机房宽度与深度的乘积。

22. 辅助机房(SECONDARY MACHINE ROOM);隔层(SECONDARY FLOOR);滑轮间隔(PULLEY ROOM)

机房在井道的上方时,机房楼板与井道顶之间的房间。它有隔间的功能,也可安装滑轮、限速器和电气设备。

23. 层站台 LANDING

各楼层用于出入轿厢的地点。

24. 层站入口 LANDING ENTRANCE

在井道壁上的开口部分,它构成从层站到轿厢之间的通道。

25. 基站 MAIN LANDING;MAIN FLOOR;HOME LANDING

轿厢无投入运行指令时停靠的层站。一般位于大厅或底层端站乘客最多的地方。

26. 预定基站 PREDETERMINED LANDING

并联或群控控制的电梯轿厢无运行指令时,指定停靠待命运行的层站。

27. 底层端站 BOTTOM TERMINAL LANDING

最低的轿厢停靠站。

28. 顶层端站 TOP TERMINAL LANDING

最高的轿厢停靠站。

29. 层间距离 FLOOR TO FLOOR DISTANCE;INTERFLOOR DISTANCE

两个相邻停靠层站层门地坎之间距离。

30. 井道 WELL;SHAFT;HOISTWAY

轿厢和对重装或（和）液压缸柱塞运动的窨。此空间以井道底坑的底井道坑的底井道壁和井道顶为界限。

31. 单梯井道 SINGLE WELL

只供一台电梯运行的井道。

32. 多梯井道 MULTIPLE WELL;COMMON WELL

可供两台或两台以上电梯运行的井道。

33. 井道壁 WELL ENCLOSURE;SHAFT WELL

用来隔开井道和其他场所的结构。

34. 井道宽度 WELL WIDTH;SHAFT WIDTH

平行于轿厢宽度方向井道壁内表面之间的水平距离。

35. 井道深度 WELL DEPTH;SHAFT DEPTH

垂直于井道宽度方向井道壁内表面之间的水平距离。

36. 底坑 PIT

底层端站地板以下的井道部分。

37. 底坑深度 PIT DEPTH

由底层端站地板至井道底坑地板之间的垂直距离。

38. 顶层高度 HEADROOM HEIGHT;TOP HEIGHT

由顶层端站地板至井道顶，板下最突出构件之间的垂直距离。

39. 井道内牛腿;加腋梁 HAUNCHED BEAM

位于各层站出入口下方井道内侧，供支撑层门地坎所用的建筑物突出部分。

40. 围井 TRUNK

船用电梯用的井道。

41. 围井出口 HATCH

在船用电梯的围井上,水平或垂直设置的门口。

42. 开锁区域 UNLOCKING ZONE

轿车厢停靠层站时在地坎上、下延伸的一段区域。当轿厢底在此区域内时,门锁方能打开,使开门机动作,驱动轿门、层门开启。

43. 平层 LEVELING

在平层区域内,使轿厢地坎与层门地坎达到同一平面的运动。

44. 平层区 LEVELING ZONE

轿厢停靠站上方和(或)下方的一段有限区域。在此区域内可以用平层装置使轿厢运行达到平层要求。

45. 开门宽度 DOOR OPENING WIDTH

轿厢门和层门完全开启的净宽。

46. 轿厢入口 CAR ENTRANCE

在轿厢壁上的开口部分,它构成从轿厢到层站之间的正常通道。

47. 轿厢入口净尺寸 CLEAR ENTRANCE TO THE CAR

轿厢到达停靠站,轿厢门完全开启后,所测得门口的宽度和高度。

48. 轿厢宽度 CAR WIDTH

平行于轿厢入口宽度的方向,在距轿厢底 1m 高处测得的轿厢壁两个内表面之间的水平距离。

49. 轿厢深度 CAR DEPTH

垂直于轿厢宽度的方向,在距轿厢底部 1m 高处测得的轿厢壁两个内表面之间水平距离。

50. 轿厢高度 CAR HEIGHT

从轿厢内部测得地板至轿厢顶部之间的垂直距离(轿厢顶灯罩和可拆卸的吊顶在此距离之内)。

51. 电梯司机 LIFT ATTENDANT

经过专门训练、有合格操作证的授权操纵电梯的人员。

52. 乘客人数 NUMBER OF PASSENGER

电梯设计限定的最多乘客量(包括司机在内)。

53. 油压缓冲器工作行程 WORKING STROKE OF OIL BUFFER

油压缓冲器柱塞端面受压后所移动的垂直距离。

54. 弹簧缓冲器工作行程 WORKING STROKE OF SPRING BUFFER

弹簧受压后变形的垂直距离。

55. 轿底间隙 BOTTOM CLEANANCES FOR CAR

当轿厢处于完全压缩缓冲器位置时,从底坑地面到安装在轿厢底下部最低构件的垂直距离(最低构件不包括导靴、滚轮、安全钳和护脚板)。

56. 轿顶间隙 TOP CLEARANCES FOR COUNTERWEIGHT

当对重装置处于完全压缩缓冲器位置时,从轿厢顶部最高部分至井道最低部分的垂直距离。

57. 对重装置顶部间隙 TOP CLEARANCES FOR COUNTERWEIGHT

当轿厢处于完全压缩缓冲器的位置时,从对重装置最高的部分至井道顶部最低部分的垂直距离。

58. 对接操作规程 DOCKING OPERATION

在特定条件下,为了方便装卸货物,货梯的轿门和层门均开启,使轿厢从底层站向上,在规定距离内以低速运行,与运载货物设备相接的操作。

59. 隔层停靠操作 SKIP-STOP OPERATION

相邻两台电梯共用一个候梯厅,其中一台电梯服务于偶数层站,而另一台电梯服务于奇数层站的操作。

60. 检修操作规程 INSPECTION OPERATION

电梯安装、维修人员必须掌握的专业技术。

61. 电梯曳引型 TRACTION TYPES OF LIFT

曳引机驱动的电梯,当机房在井道上方的为顶部曳引型;当机房在井道侧面的为侧面曳引型。

62. 电梯曳引机绳曳引 HOIST ROPES RATIO OF LIFT

悬吊轿厢的钢丝绳根数与曳引轮单侧的钢丝绳根数之比。

63. 消防服务 FIREMAN SERVICE

操纵消防开关能使电梯投入消防员专用的状态。

64. 独立操作 INDEPENDENT OPERATION

靠钥匙开关来操控轿厢内按钮使轿厢升降运行。

65. 缓冲器 BUFFER

用来吸收轿厢动能的一种弹性缓冲安全装置。

66. 油压缓冲器;耗能型缓冲器 HYDRAULIC BUFFER;OIL BUFFER

以油为介质来吸收轿厢或对重产生动能的缓冲器。

67. 弹簧缓冲器；蓄能型缓冲器具 SPRING BUFFER

以弹簧变形来吸收轿厢或对重产生动能的缓冲器。

68. 减振器具 VIBRATING ABSORBER

用来减小电梯运行所产生的振动和噪声的装置。

69. 轿厢 CAR；LIFT CAR

运载乘客或其他载荷的轿体部件。

70. 轿厢底；轿底 CAR PLATFORM；PLATFORM

在轿厢底部，支撑载荷的组件，它包括地板、框架等。

71. 轿厢壁；轿壁 CAR ENCLOSURES；CAR WALLS

由金属板与轿厢底、轿厢顶和轿厢门围成的一个封闭空间。

72. 轿车厢顶；轿顶 CAR ROOF

在轿厢的上部，具有一定强度要求的顶盖。

73. 轿厢扶手 CAR HANDRAIL

固定在轿厢壁上的扶手。

74. 轿顶防护栏杆 CAR PROTECTION BALUSTADE

设置在轿顶上部，对维修人员起防护作用的构件。

75. 轿厢架；轿架 CAR FRAME

固定和支撑桥厢的框架。

76. 检修门 ACCESS DOOR

开设在井道壁上，通向底坑或滑轮间供检修人员使用的门。

77. 手动门 MANUALLY OPERATED DOOR

用人力开关的轿门或层门。

78. 自动门 POWER OPERATED DOOR

靠动力开关的轿门或层门。

79. 层门;厅门 LANDING DOOR;SHAFT DOOR;HALL DOOR

设置在层站入口的门。

80. 防火层门;防火门 FIRE-PROOF DOOR

能防止或延缓炽热气体或火焰通过的一种层门。

81. 轿厢门;轿门 CAR DOOR

设置在轿厢入口的门。

82. 安全触板 SAFETY EDGES FOR DOOR

在轿厢门关闭过程中,当有乘客或障碍物触及时,轿门重新打开的机械门保护装置。

83. 铰链门;外敞门 HINGED DOOR

门的一侧为铰链联系,由井道向通道方向开启的层门。

84. 栅栏门 COLLAPSIBLE DOOR

可以折叠,关闭后成栅栏形状的轿厢门。

85. 水平滑动门 HORIZONTALLY SLIDING DOOR

沿门导轨和地坎槽水平滑动开启的层门。

86. 中分门 CENTER OPENING DOOR

层门或轿门由门口中间各自向左、右以相同速度开启的门。

87. 旁开门；双折门；双速门 TWO-SPEED SLIDING DOOR；TWO-PANEL SLIDING DOOR；TWO SPEED DOOR

层门或轿门的两扇门以两种不同速度向同一侧开启的门。

88. 左开门 LEFT HAND TWO SPEED SLIDING DOOR

面对轿厢，向左方向开启的层门或轿门。

89. 右开门 RIGHT HAND TWO SPEED SLIDING DOOR

面对轿厢，向右方向开启的层门或轿门。

90. 垂直滑动门 VERTICALLY SLIDING DOOR

沿门两侧垂直门导轨滑动开启的门。

91. 垂直中分门 BIPARTING DOOR

层门或轿门的两扇门由门中间以相同速度各自向上、下开启的门。

92. 曳引绳补偿装置 COMPENSATING DEVICE FOR HOIST ROPES

用来平衡由于电梯提升高度过高、曳引绳过长造成运行过程中偏重现象的部件。

93. 补偿绳装置 COMPENSATING ROPE DEVICE

用钢丝绳及张紧轮构成的补偿装置。

94. 补偿绳防跳装置 ANTI-REBOUND OF COMPENSATION ROPE DEVICE

当补偿绳张紧装置超出限定位置时，能使曳引机停止运转的电气安全装置。

95. 地坎 SILL

位于电梯轿厢或层门入口处的带槽金属踏板。

96. 轿厢地坎 CAR SILLS

轿厢入口处的地坎。

Chapter 6　Troubleshooting and Maintenance

97. 层门地坎 LANDING SILLS
层门入口处的地坎。

98. 轿顶检修装置 INSPECTION DEVICE ON TOP OF THE CAR
设置在轿顶上部,供检修人员检修时应用的装置。

99. 轿顶照明装置 CAR TOP LIGHT
设置在轿顶上部,供检修人员检修时照明的装置。

100. 底坑检修照明装置 LIGHT DEVICE OF PIT INSPECTION
设置在井道底坑,供检修人员检修时照明的装置。

101. 轿厢内指层灯;轿厢位置指示 CAR POSITION INDICATOR
设置在轿厢内,显示其运行层站的装置。

102. 层门门套 LANDING DOOR JAMB
装饰层门门框的构件。

103. 层门指示灯 LANDING INDICATOR;HALL POSITION INDICATOR
设置在层门上方或一侧,显示轿厢运行层站和方向的装置。

104. 层门方向指示灯 LANDING DIRECTION INDICATOR
设置在层门上方或一侧,显示轿厢运行方向的装置。

105. 控制屏 CONTROL PANEL
有独立的支架,支架上有金属绝缘底板或横梁,各种电子器件和电器元件安装在底板或横梁上的一种屏式电控设备。

106. 控制柜 CONTROL CABINET;CONTROLLER
各种电子器件和电器元件安装在一个有防护作用的柜形结构内的电控设备。

107. 操纵箱；操纵盘 OPERATION PANEL;CAR OPERATION PANEL

用于操纵轿厢运行的电气装置。

108. 警铃按钮 ALARM BUTTON

电梯内的"报警电话"或"警铃"，用来报警和提醒管理人员以获得救助。

4C: Evaluation. 回顾所学知识，写下你已掌握的单词和句子。

Write down the words that you think are the most useful:

1. _____
2. _____
3. _____
4. _____
5. _____
6. _____
7. _____
8. _____

List the sentences that you think are the most important:

1. _____
2. _____
3. _____
4. _____
5. _____
6. _____
7. _____
8. _____

Appendix 1　International Phonetics Alphabet
附录一　英语国际音标表(48个)

国际音标表

元音	单元音	长元音	/iː/ /uː/ /ɔː/ /ɑː/ /ɜː/
		短元音	/ɪ/ /ʊ/ /ɒ/ /ʌ/ /ə/ /æ/ /e/
	双元音		/eɪ/ /aɪ/ /ɔɪ/ /aʊ/ /əʊ/ /ʊə/ /ɪə/ /eə/
辅音	清浊成对		/p/ /t/ /k/ /f/ /s/ /θ/ /tʃ/ /ʃ/ /tr/ /ts/
			/b/ /d/ /g/ /v/ /z/ /ð/ /dʒ/ /ʒ/ /dr/ /dz/
	其他		/w/ /j/ /l/ /h/ /r/ /m/ /n/ /ŋ/

Appendix 2　Keys to Exercises
附录二　练习参考答案

Chapter 1　Section C

Fill in the form. 填写表格。

Meeting in an Elevator of the Company	
Zhang Ming's floor	Floor 28
Judy's floor	Floor 25
Zhang Ming's work	Good busy
Judy's work	Fine
Zhang and Judy's relationship	Workmates

Answer the following questions. 回答下列问题。

1. Elisha Otis is an American.
2. He invented the first safety brake for elevator.
3. It stops the lift falling if the cords are broken.
4. Yes, it did.

Chapter 2　Section C

Decide the sentences T (true) or F (false). 判断句子正误。

1. F　2. T　3. F　4. T　5. T

Fill in the form. 填写表格。

A Freight Elevator	
Function	to carry freight or goods
Capacity	larger and greater
Loads	from 2,500kg to 4,500kg, and some even 100,000 pounds (45,359kg)
Door	a manual door, and sometimes multiple doors

Chapter 3　Section C

Fill in the form. 填写表格。

Parts	Function
Cars metal boxes	raise up and down
Counterweights	balance the cars
Electric motor	hoist the cars up and down
Metal cables and pulleys	run between the cars and the motors
Safety systems	protect the passengers if a cable breaks

Decide the sentences T (true) or F (false). 判断句子正误。

1. T　2. F　3. F　4. F　5. F

Chapter 4   Section C

Answer the following questions. 回答以下问题。

1. They may work in narrow spaces or awkward positions.
2. The dangers include falls, electrical shock, muscle strains, and other injuries related to handling heavy equipment.
3. Workers often wear hardhats, harnesses, ear plugs, safety glasses, protective clothing, shoes and masks.
4. Yes, it does.

Decide the sentences true (T) or false (F). 判断句子正误。

1. T   2. F   3. F   4. F   5. T

Chapter 5   Section B

Translate the following sentences. 翻译下列句子,注意画线部分词语。

1. 请遵循说明书安装和使用电梯。
2. 电梯警铃应该足够响,听得到。
3. 电梯的检测和验收包括电梯安装和功能的检测和验收。
4. 客货双用电梯是可运载乘客和货物的电梯。
5. 电子设备包括接地、安全设备等。

附录二 | 练习参考答案

Section C

Answer the following questions. 回答以下问题。

1. Such as balance test, full load test, speed test, temperature rise test, car leveling test, brake test, locking test and so on.

2.
Number	Meaning
3	Floor leveling accuracy of any landing floor should be within plus or minus 3 mm.
125	The brake system examination and test should be carried out when the car is going down at rated speed with 125% of the rated load and interrupting the supply to the motor and the brake.
7	The car should not be able to start until the locking elements are engaged by at least 7mm.

Answer the following questions. 回答以下问题。

1. The world's tallest elevator test tower will be in Shanghai.
2. It will be finished by the end of 2018.
3. Its height will be around 270 meters.
4. Philippe Delpech is the Otis President.
5. It is designed to test the stress and limits of special elevator cars. Tests are also carried out in the tower to insure reliability and safety in elevator designs and solve any problems.

Chapter 6 Section B

Translate the following sentences. 翻译下面的句子。

1. 检查层门和厅门是关上的,没有什么限制轿厢运行。

2. 轿厢没有接近平层时,不要试图移动失速的电梯或移动乘客。

3. 当按钮松开时,灯不会亮。

4. 关于发动机负载,正常电流状态下,延时熔断保险丝应该在100%到125%的负载在正常运行。

5. 过分的轿厢振动是因为电力(驱动系统)或操作系统的机械问题。

Section C

Answer the following questions. 回答下列问题。

1. The lift she was in stopped between the 11th and the 12th floor.
2. Yes, she was.
3. She followed the instructions in the lift and pushed the emergency button.
4. The technician told Joan to stay calm and relax, while they tried to repair it.
5. About 20 minutes.
6. He thought he was a rude guy.

Answer the following questions. 回答下列问题。

1. Remain calm, call for help and deal with loneliness.
2. No, there aren't.
3. It is very likely that the worst thing is that you'll have to wait a short while until the lift is working again.
4. Try pushing a floor button again, or the open door button.
5. There is a telephone or an alarm button to call for help, use a mobile phone, shout loudly or bang on the door to let someone hear outside.
6. You may want someone to be with you.

Appendix 3　Vocabulary
附录三　词汇表

Chapter 1

elevator /ˈelɪveɪtə/ n. 电梯；升降机
escalator /ˈeskəleɪtə/ n. 自动扶梯
low speed elevator 低速电梯
high speed elevator 高速电梯
super high speed elevator 超高速电梯
in case of danger 危急情况
stair /steə/ n. 楼梯
brake /breɪk/ n. 刹车
capacity /kəˈpæsəti/ n. 载容量
top floor 顶楼
rated capacity 额定载容量
Elisha Graves Otis 伊莱沙·格雷夫斯·奥的斯
be full of 充满
company /ˈkʌmpəni/ n. 公司
upwards /ˈʌpwədz/ adv. 向上
pulley /ˈpʊli/ n. 滑轮；滑车
attach /əˈtætʃ/ v. 贴；装；连接
cord /kɔːd/ n. 粗线；绳
invention /ɪnˈvenʃn/ n. 发明

lift /lɪft/ n. 电梯；升降机
lean /liːn/ v. 斜靠
mid-speed elevator 中速电梯
express lift 高速电梯
take an elevator 乘坐电梯
operate /ˈɒpəreɪt/ v. 操作；运转
skyscraper /ˈskaɪskreɪpə/ n. 摩天大楼
rate /reɪt/ n. 比率
safety brake 紧急刹车，安全制动
start off 开始

staff /stɑːf/ n. 全体职员
hold the door 别关门
technology /tekˈnɒlədʒi/ n. 科技
system /ˈsɪstəm/ n. 系统；制度
fall /fɔːl/ v. 落下；跌落
confidence /ˈkɒnfɪdəns/ n. 信心
pulley system 滑轮系统

The Egyptian Pyramids 埃及金字塔
steam engine 蒸汽发动机
magic /ˈmædʒɪk/ adj. 有魔力的
cliff /klɪf/ n. 悬崖，峭壁
uncertain /ʌnˈsɜːtn/ adj. 不确定的
environment /ɪnˈvaɪrənmənt/ n. 环境
exposure /ɪkˈspəʊʒə/ n. 暴露
full-exposure 全暴露的
outdoor elevator 户外电梯
Guinness World Record 吉尼斯世界纪录

Chapter 2

freight /freɪt/ n.（海运、空运或陆运的）货物
sightseeing /ˈsaɪtsiːɪŋ/ n. 游览；观光
motor /ˈməʊtə/ n. 汽车；发动机
residential /ˌrezɪˈdenʃl/ adj. 适合居住的
dragger /ˈdrægə/ n. 曳引机
buffer /ˈbʌfə/ n. 缓冲器
safety /ˈseɪfti/ n. 安全；平安
governor /ˈgʌvənə/ n. 限速器
passenger elevator 乘客电梯
electrical control cabinet 电子控制柜
call box 呼梯盒
floor gate 层门
safety jaw 安全钳
conveyor belt 传送带
speed reducer 减速器
counter unit 对重装置
floor indicator 楼层指示器
elevator cable 电梯电缆
relax /rɪˈlæks/ v. 放松
develop /dɪˈveləp/ v. 发展
design /dɪˈzaɪn/ v. 设计
vehicle /ˈviːəkl/ n. 交通工具；车辆
load /ləʊd/ n. 负载；载重量
provide...in use 投入使用
parking garage 停车场
residential building 住宅楼
Boeing 747 波音747飞机
passenger double-deck aircraft lift 乘客双层飞机电梯
aside /əˈsaɪd/ adv. 在旁边
button /ˈbʌtn/ n. 按钮；纽扣
nervous /ˈnɜːvəs/ adj. 紧张不安的
calm /kɑːm/ adj. 平静的；镇静的
uncomfortable /ʌnˈkʌmftəbl/ adj. 使人不舒服的
call buttons 呼梯按钮
stare at 盯着看
in the case of an emergency 紧急情况时
follow the instructions 按照指令
goods /gʊdz/ n. 货物
notice /ˈnəʊtɪs/ n. 通知
allow /əˈlaʊ/ v. 允许；许可
operator /ˈɒpəreɪtə/ n. 操作者

though /ðəʊ/ conj. 虽然；尽管
manual /ˈmænjuəl/ adj. 手动的
pound /paʊnd/ n. 英镑
arch /ɑːtʃ/ n. 拱门
link /lɪŋk/ v. 连接；联系
foot /fʊt/ n. 英尺（复数 feet）
magnificent /mæɡˈnɪfɪsnt/ adj. 壮丽的；宏伟的
must see 必看的
St. Louis, Missouri (美国)密苏里州圣路易斯市

handle /ˈhændl/ v. 操纵；搬运
multiple /ˈmʌltɪpl/ adj. 多种多样的
gateway /ˈɡeɪtweɪ/ n. 门
symbol /ˈsɪmbl/ n. 象征
height /haɪt/ n. 高度
structure /ˈstrʌktʃə/ n. 结构；构造

Chapter 3

counterweight /ˈkaʊntəweɪt/ n. 对重
sensor /ˈsensə/ n. 传感器
motor /ˈməʊtə/ n. 发动机；马达，汽车
wire rope 钢丝绳
electric motor 电动机
rotational speed 转速
elevator well 电梯机井
electronic control system 电气控制系统
door opener system 开门机系统
inner door 轿门
hoist /hɔɪst/ v. 升起；吊起
ensure /ɪnˈʃɔː; -ˈʃʊə/ v. 保证；担保
electric motor 电马达
hooked up 钩住；提拉
protect from doing sth. 保护……以免……
viewpoint /ˈvjuːpɔɪnt/ n. 观点；看法
lower /ˈləʊə/ v. 使……降下
still /stɪl/ adj. 静止的；寂静的

shaft /ʃɑːft/ n. 机井
balance /ˈbæləns/ n. 平衡
rail /reɪl/ n. 轨道
guide shoe 导靴
brake system 制动器系统
safety system 安全系统
elevator computer 电梯控制板
elevator mechanical system 电梯机械系统
outer door 层门，厅门
door rail 门导轨
factor /ˈfæktə/ n. 因素；要素
safety device 安全设备
cut off power 切断电源
motion sensor system 运行传感器系统

raise /reɪz/ v. 提升
see-saw /ˈsiːsɔː/ n. 跷跷板
location /ləʊˈkeɪʃn/ n. 地点；位置

release /rɪˈliːs/ v. 松开,释放　　grab /græb/ v. 抓住;夺取
hook /hʊk/ v. (使)钩住; n. 钩子　　casino /kəˈsiːnəʊ/ n. 赌场
version /ˈvɜːʃn/ n. 版本　　generic /dʒəˈnerɪk/ adj. 通用的
Luxor Hotel 卢克索酒店
Luxor Inclinator Elevator 卢克索倾斜仪电梯
39 degree angle 39 度角　　Las Vegas 拉斯维加斯(美国城市名)
Nevada 内华达州(美国州名)

Chapter 4

breaker /ˈbreɪkə/ n. 断路器　　pit /pɪt/ n. 底坑
installer /ɪnˈstɔːlə/ n. 安装员　　penthouse /ˈpenthaʊs/ n. 机房;阁楼套房
selector /sɪˈlektə/ n. 选层器　　engineer /ˌendʒɪˈnɪə/ n. 工程师
repairer /rɪˈpeərə/ n. 修理员　　speed reducer 减速器
lift motor room 机房　　contactor /ˈkɒntæktə/ n. 触点
relay /ˈriːleɪ/ n. 继电器;中继设备　　resistor /rɪˈzɪstə/ n. 电阻器
transformer /trænsˈfɔːmə/ n. 变压器　　installation /ˌɪnstəˈleɪʃn/ n. 安装;设置
drill /drɪl/ n. 钻;钻机　　waterproof /ˈwɔːtəpruːf/ adj. 防水的
maintenance /ˈmeɪntənəns/ n. 维护　　landing call 层站停靠
car call 轿厢呼叫　　major in 主攻,专业
home landing parking 基站停靠　　MRL installation 无机房安装
awkward /ˈɔːkwəd/ adj. 令人尴尬的　　injury /ˈɪndʒəri/ n. 伤害;损害
hardhat /ˈhɑːdhæt/ n. 安全帽　　harness /ˈhɑːnɪs/ n. 背带,保护带
plug /plʌg/ n. 塞子;插头　　mask /mɑːsk/ n. 面具
electrical shock 电击　　muscle strain 肌肉拉伤
relate to 与……相关　　protective clothing 防护服
construction trade 建筑行业　　visible /ˈvɪzəbl/ adj. 看得见的
energy efficient 节约能源　　storey /ˈstɔːri/ n. 楼层
descent /dɪˈsent/ n. 下降　　push the limit 推向极限
peak speed 巅峰速度

Chapter 5

handrail /ˈhændreɪl/ *n.* 扶手
buffer /ˈbʌfə/ *n.* 缓冲器
landing /ˈlændɪŋ/ *n.* 层站；楼梯平台
deliver /dɪˈlɪvə/ *v.* 输送
engage /ɪnˈɡeɪdʒ/ *v.* 咬合，(使)衔接
specification /ˌspesɪfɪˈkeɪʃn/ *n.* 说明书
earthing /ˈɜːθɪŋ/ *n.* 接地
safety gear 安全钳
leveling accuracy 平层精度
locking element 锁紧元件
run /rʌn/ *v.* 运行
detail /ˈdiːteɪl/ *n.* 细节
plus /plʌs/ *adj.* 零上；好的；*prep.* 加
put into service 投入使用
temperature rise test 升温测试
giant /ˈdʒaɪənt/ *n.* 巨人
facility /fəˈsɪləti/ *n.* 设施；设备
industry /ˈɪndəstri/ *n.* 工业；行业
stress /stres/ *n.* 压力
reliability /rɪˌlaɪəˈbɪləti/ *n.* 可靠性
keep pace with 跟上步伐
Philippe Delpech 菲利普戴培杰（奥的斯电梯全球总裁）
originally /əˈrɪdʒənəli/ *adj.* 起初
booth /buːð; NAmE buːθ/ *n.* 不受干扰的划定空间（如电话亭、投票间等）
interior /ɪnˈtɪəriə/ *adj.* 在内的；内部的
helicoidal /ˌhelɪˈkɔɪdəl/ *adj.* 螺旋状的
Lisbon /ˈlɪzbən/ 里斯本（葡萄牙首都）

clamp /klæmp/ *n.* 夹钳；*v.* 夹住
governor /ˈɡʌvənə/ *n.* 限速器
spark /spɑːk/ *v.* 产生火花；*n.* 火花
accept /əkˈsept/ *v.* 验收
strike /straɪk/ *v.* 碰；撞击
audible /ˈɔːdəbl/ *adj.* 听得见的
examine and test 检测
guide rail 导轨
rated load 额定负载
equipment /ɪˈkwɪpmənt/ *n.* 装备；设备
fully load 满员
interrupt /ˌɪntəˈrʌpt/ *v.* 使中断；插嘴
minus /ˈmaɪnəs/ *adj.* 负的；小于零的
carry out 执行
regardless of 无视，不管
Otis 奥的斯电梯公司
global /ˈɡləʊbl/ *adj.* 全球的
structure /ˈstrʌktʃə/ *n.* 结构
limit /ˈlɪmɪt/ *n.* 限度
elevator test tower 电梯测试塔

urban /ˈɜːbən/ *adj.* 城市的
vertical /ˈvɜːtɪkl/ *adj.* 垂直的
uphill /ˌʌpˈhɪl/ *adj.* 上坡的

Carmo Square 卡莫广场
electrical operation 电气操作
neogothic style 新哥特式风格
Santa Justa Lift 圣胡斯塔升降机

Chapter 6

fuse /fjuːz/ v. （使）熔化；n. 保险丝
bearing /ˈbeərɪŋ/ n. 轴承
lead /liːd/ n. 电线
jump /dʒʌmp/ n. 短接
stall /stɔːl/ adj. 失控的
blown /bləʊn/ adj. 熔断的
coasting /ˈkəʊstɪŋ/ n. 溜梯
rotary encoder 旋转编码器
balance factor 平衡系数图
traction sheave 曳引轮
short circuit 短路
vibrate /vaɪˈbreɪt/ v. 震动
rating current 额定电流
security contact 安全触点
photo eye 光电检测器
troubleshooting /ˈtrʌblʃuːtɪŋ/ n. 故障
failure /ˈfeɪljə/ n. 失败
emergency /iˈmɜːdʒənsi/ n. 紧急情况
reopen /ˌriːˈəʊpən/ v. 再打开，重新打开
occasionally /əˈkeɪʒnəli/ adv. 偶然
maintenance /ˈmeɪntənəns/ n. 保养；维护
blindly /ˈblaɪndli/ adv. 盲目地
verify /ˈverɪfaɪ/ v. 证实；查对
excessive /ɪkˈsesɪv/ adj. 过分的
anxious /ˈæŋkʃəs/ adj. 忧虑的；焦急的
technician /tekˈnɪʃn/ n. 技术员
calm /kɑːm/ adj. 平静的，镇静的
push sb. aside 把某人推到一边
stuck /stʌk/ adj. stick 的过去式（分词），被卡住的；动不了的
plummet /ˈplʌmɪt/ v. 垂直落下，骤然下跌
patiently /ˈpeɪʃntli/ adv. 耐心地
panic /ˈpænɪk/ adj. 惊恐
trap /træp/ v. 使陷入绝境
alarm button 警示按钮
emergency services 紧急服务
Switzerland /ˈswɪtsələnd/ n. 瑞士
exterior /ɪkˈstɪəriə/ adj. 外部的
plateau /ˈplætəʊ/ n. 高原
overlook /ˌəʊvəˈlʊk/ v. 俯视
summit /ˈsʌmɪt/ n. 峰会
upgrade /ʌpˈɡreɪd/ v. 升级
Burgenstock 比尔根山
Hammetschwand Elevator 哈梅茨施万德观光电梯
Lake Lucerne 琉森湖